REACTION KINETICS FOR A-LEVEL CHEMISTRY

REACTION KINETICS

for

A-LEVEL CHEMISTRY

… and a bridge to undergraduate study

Mark Burkitt

Westcott Research and Consulting

Published in 2021

by

Westcott Research and Consulting

ACKNOWLEDGEMENTS

I should like to thank my family for the time and space they have allowed me to devote to the writing of this book, which is dedicated to them. My wife, Dr Louise Race, deserves a special thanks because she was burdened with the task of proofreading; given that she hung up her lab coat several years ago, this cannot have been an easy task, yet she also made many helpful comments and suggestions. In the time-honoured fashion, however, I assume full responsibility for any remaining errors. I would also like to thank my daughter, Hebe, for her help in designing the cover.

I thank especially Professor Patrick Riley, whom I first met when I was still 'doing' kinetics in the laboratory. Through our many stimulating discussions and emails, he has supported me with enthusiasm and valued feedback. My thanks are also due to the many individuals who, over the years, have enhanced my understanding of reaction kinetics. Notable in this regard are Professors Bruce Gilbert and Peter Wardman.

Although I was *their* teacher, the numerous students I have taught over the past 14 years have, by forcing me to constantly evaluate and review how I explain reaction kinetics, deepened my own understanding of the subject. I am particularly grateful to these students for convincing me of the importance of teaching challenging material in the depth it demands. Invariably, I find young people appreciate being credited with the ability to follow complicated reasoning.

None of the graphs shown in the figures are reproductions. They were all simulated using physical constants published in the scientific literature. I am grateful to the authors of the published works from which I have obtained the relevant values. All my original sources have been referenced in the 'Further reading' section and, where a study has been discussed in some detail, also in the main text. Simulations were carried using Microsoft Excel. Other Microsoft programs (Word and PowerPoint) were also used in the production of this book.

Mark Burkitt
October 2021

CONTENTS

Chapter 3

DERIVING REACTION MECHANISMS FROM RATE EQUATIONS: 'PUTTING THE HORSE BEFORE THE CART'

Chapter 4

REACTION KINETICS IN THE SCHOOL LABORATORY: METHODS OF INVESTIGATION

Appendix I
HOW TO CONSTRUCT AND COMBINE HALF-EQUATIONS

PREFACE

As a private tutor, one of the most rewarding moments for me is when a student – who initially embarked upon A-level chemistry because it is a requirement for entry to the of study medicine, dentistry or one of the increasingly popular biomedical science courses at university – tells me (towards the end of their studies) that chemistry is their favourite subject. I have even had students studying chemistry as their only science A-level, intent on reading humanities at university, decide to switch to a career in chemistry. These students all give the same reason for their love of chemistry (their words, not mine): the account of the natural world given by chemistry is hugely intellectually satisfying.

Chemistry is, of course, based on physics. The descriptions of nature embodied in, for example, thermodynamics (energetics) and the Boltzmann distribution are rooted in physics, but it is how they are then used to make sense of chemistry that I believe students find so satisfying. In reality, the transition between physics and chemistry is a seamless one, as is that into the biological sciences.

Intellectual satisfaction comes only with depth of knowledge and genuine understanding. That is why it is worth reading a whole book on a single chemistry topic, especially a topic so important and enjoyable as reaction kinetics – the study of the rates of chemical reactions. The few 'spreads' devoted to kinetics in the standard A-level textbooks do not give you the depth of understanding and breadth of knowledge needed to really understand the topic, let alone tackle the more challenging examination questions. Of course, the purpose of the course textbooks is, essentially, only to support what you are taught in class – it is assumed the teacher will expand on the material and explain the various nuances of the topic. Think about how long a textbook would be if your teacher were to write down everything said to you over the course of the two or three double periods typically spent on reaction kinetics – it would be at least as long as this book.

Another reason for the length of this book is the style in which the material is presented. Unlike a textbook intended mainly for classroom use, there is unlikely to be a teacher to hand should you not understand a section (the school holidays would be a good time to read this book), therefore every effort has been made to explain the material as clearly and in as simple terms as possible. Do not, however, be worried if you do not understand everything. Although it is recommended you read the chapters in numerical order, the key concepts are revisited within different contexts, so if one explanation leaves you confused, you may well understand it when it reemerges later. There is also a handy summary of the key points covered at the end of each chapter.

I have made a particular effort to illustrate the various phenomena using real, fully worked examples, for which I have researched the scientific literature to find the appropriate rate constants, activation energies and Arrhenius constants, as well as the most widely agreed reaction mechanisms. I believe this approach to be more conducive to learning than resorting to the generic 'A reacts with B to form C and D', with made-up physical constants. I cannot emphasise too much the benefits to be gained from working through the worked examples: Plot the data from the tables as I have done; measure the gradients; rearrange the equations; and check my calculations, which have been presented assuming you have only the most rudimentary skills in mathematics, with no knowledge of Euler's number (e) and natural logarithms (ln). No matter how straightforward these tasks may seem when someone else (yours truly) does them for you, you will be surprised by how many mistakes you make when doing it all yourself for the first time – do not discover this truth in the examinations!

Reaction kinetics, I would argue, is the most difficult topic in the A-level chemistry courses. Many students believe kinetics to be just a bit of maths – solving one or two equations – and would say that electrochemistry or NMR, for example, are more difficult, but they are not. Provided you have been taught these topics properly, the examination questions are relatively straightforward: there is only a limited number of ways in which the examiners can test your understanding. The same is true of energetics, transition metals and (to some extent) organic chemistry.

Although many questions on kinetics *begin* in much the same way (derive a rate equation and rate constant – or use these to find the missing values in a table of data), they often then ask you to suggest a mechanism for a reaction you may not have seen before. Unless you have a very solid understanding of the principles of reaction kinetics and a broad knowledge of reaction mechanisms, this can be where you come unstuck. In their final lessons before their examinations, I try to run my students through as many of the reaction mechanisms – and their associated rate equations – that have appeared in previous papers as possible. These include first- and second-order nucleophilic substitutions, the Harcourt-Essen reaction and the iodination of propanone. If you are already familiar with the kinetics and mechanisms of such reactions, it is relatively easy to see where an examination question is going than to have to figure it all out for yourself – especially when asked to suggest a mechanism. No amount of preparation can cover all the reactions that may come up in the examinations, but the more reactions you have studied in detail, the more likely you will be able to apply the principles to an unfamiliar reaction. You will acquire much of the required depth and breadth of knowledge from reading this book. Experience in the application of the material to examination questions can be gained by reviewing the questions I have described individually in Appendix III.

Another reason why kinetics can be a particularly difficult topic concerns the wide range of techniques that can be used to measure the rate of a reaction. In some questions, the rates have been calculated for you, but in others you may be required to derive these by drawing tangents or by processing raw data – for example, from a titration or one of the 'clock' methods. This is why it is so important for students to do lots of practical work and invest the time needed to research the methods they are using, process their results and research the answers to the questions accompanying each practical. Very often, one finds the material in an examination question covered in a Required Practical.

Much as every effort has been made to make reaction kinetics as easy as possible to understand (using detailed explanations and fully worked examples), this book is intended to take students' understanding of the subject at A-level to the very highest level. The content will stretch students more than the standard A-level textbooks (and may be a useful resource for teachers wishing to challenge their most able students), but if further justification for this is needed, the student need only look at one or two questions in the new (post-2015) A-level examinations to see the extent to which the content in the specifications is being stretched to identify the top candidates (see, for example, the questions in Appendix III). Nothing covered in this book is not relevant at A-level. I have explained one or two topics that are not addressed directly – by name - in the specifications (*e.g.* the use of pseudo first-order reaction conditions and the Lindemann model of first-order reactions), but the rationale and concepts I have used in their explanation are largely accessible through the application of A-level material. Admittedly, the Lindemann model is stretching things a little, which is why it is covered in an appendix. I would, however, add that, at the time of writing, with so many students being awarded A*/A grades, there is talk of the universities setting their own entrance examinations. Make no mistake about it, these examinations will be tougher than the A-levels.

Should you be planning to study chemistry or a course that includes a significant component of chemistry (*e.g.* biochemistry or biomedical sciences) at university, this book should help you bridge the gap between study at A-level and that as a first-year undergraduate. The main difference between how reaction kinetics is covered in introductory university courses and at A-level is in the mathematics, where at university you will be using integrated rate equations. These allow you to calculate the concentration of a reactant (or product) at any chosen time. At A-level, you are dealing primarily with the *rates* at which these concentrations change. Integrated rate equations were used to calculate the concentration-time plots shown in this book. Students reading this book in conjunction with their university studies will find the references to the some of the original scientific papers given in the footnotes and in the 'Further reading' section useful. Those whose studies are confined to A-level need not be distracted by such matters, but should review the selection of examination questions given in Appendix III.

University students (and indeed chemistry teachers) will notice very quickly that, in aiming to pitch the material appropriately at A-level standard, gross simplifications have been made. I have, for example, incorporated concepts from transition-state theory into my description of collision theory. Similarly, no distinction has been made between diffusion- and activation-controlled reactions. These are details which, if addressed in this volume, would negate its primary purpose as a book for study at A-level. Finally, notwithstanding my constant reference to the A-level examinations, it is hoped this book will be useful to those studying for other, equivalent-level examinations, including Scottish Highers and the International Baccalaureate (Higher).

Mark Burkitt
October 2021

Chapter 1

FROM THE RATE CONCEPT TO THE RATE EQUATION

Two motorists have each completed separate journeys, one covering 20 km, the other 225 km. Which was the faster? To answer this question it is, of course, necessary to know the time taken for each journey – speed being distance divided by time. Assuming the 20 km journey took 15 minutes and the 225 km one 5 hours, then the calculated speeds are, respectively, 80 and 45 km/h.

We all know intuitively that it cannot be assumed that the driver covering the greater distance was the faster, yet students commonly fail to apply similar reasoning when considering the 'speeds' or *rates* of chemical reactions. The rate of a reaction is – not withstanding important caveats that need not concern us here – the chemical equivalent to the speed of a car (or any other moving object). Chemical reactions do not 'go to places', so to speak, but it can be helpful to consider the progress of a reaction as a journey from reactants to products. Indeed, in considering if a proposed reaction is possible, chemists often speak in terms of whether or not it 'goes'. Similarly, we often hear of the mechanism of a chemical reaction described in terms of the 'route' from reactants to products.

1.1 Determining the rate of a reaction – measuring its speed over an ever-diminishing time interval

The extent to which a chemical reaction has taken place at a chosen time-point can be stated in terms of how much of the starting material (reactant) has been converted into product. Consider, for example, the nucleophilic substitution reaction that occurs when sodium hydroxide solution is added to bromoethane:

$$CH_3CH_2Br + OH^- \rightarrow CH_3CH_2OH + Br^-$$

The progress of this reaction can be conveniently followed by monitoring the consumption of hydroxide ions, measured by titrating small samples of the reaction mixture removed at chosen time-points. The graph below (**Figure 1.1**) shows how the concentration of hydroxide ions decreases over 4 hours at 50 °C (this reaction is not particularly fast).

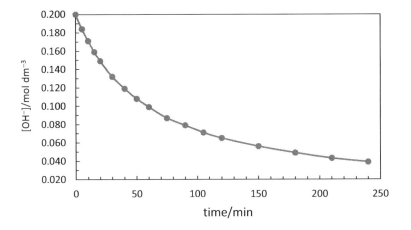

Figure 1.1   Time-course of hydroxide ion concentration during the reaction of sodium hydroxide with bromoethane at 50 °C. The starting concentration of each reactant was 0.200 mol dm^{-3}.

Over the course of 240 minutes, the concentration of the OH$^-$ ion has fallen by 0.161 mol dm^{-3} (from its initial value of 0.200 mol dm^{-3} to 0.039 mol dm^{-3}). Dividing this change by the time over which it has taken place gives us the rate of reaction:

$$\text{rate} \quad = \quad \frac{0.161 \ \text{mol dm}^{-3}}{240 \ \text{min}}$$

$$= \quad 6.71 \times 10^{-4} \quad \frac{\text{mol dm}^{-3}}{\text{min}}$$

Notice how the units are shown in the form of the fraction in which they were entered into the calculation (mol dm^{-3} 'over' min). Any unit can be moved between the 'top' and 'bottom' of a fraction by reversing the sign of its power. A power of +1 is not usually shown, therefore the 'mol' and 'min' units in the above expression represent mol^1 and min^1. If we wish to move the 'min' to the top, thereby placing all the units neatly on a single line, we must indicate that this has been done by writing its power as −1:

$$\frac{mol\ dm^{-3}}{min} = mol\ dm^{-3}\ min^{-1}$$

You may have noticed the unit of volume (dm^3) has also been given a negative power. This is because the units of concentration arise 'spontaneously' as $mol\ dm^{-3}$ when we divide the amount of a substance (in moles) by the volume in which it is contained:

$$concentration = \frac{amount}{volume} = \frac{mol}{dm^3} = mol\ dm^{-3}$$

Although you may be tempted to simply memorise the various units used in chemistry, not worrying how they 'emerge' spontaneously from calculations, this is not the way to build the versatility and confidence you will need to be able to cope with the extended calculations you are likely to encounter in the examinations. Examination questions are designed to test how well students *understand* what they are doing: their purpose is not to see whether they can plug numbers into formulae, without knowing what they are doing, and then plonk on at the end a set of memorised, off-the-shelf units. It is very important, therefore, that you cultivate the 'good habit' of working with the units in your calculations: practise tracing the units associated with the numbers you enter into calculations (mol, dm^3, min, s *etc*) all the way through to those in your final answer, allowing for the cross-multiplication and cancelling down steps involved. Deriving the units of rate for the reaction of sodium hydroxide with bromoethane, as we did above, is not particularly taxing; however, the basic principles involved are just the same in the more complicated calculations you can expect to encounter in your examinations.

Think also about what your calculations are telling you about a chemical reaction: the rate we have calculated above for the reaction of sodium hydroxide with bromoethane, $6.7 \times 10^{-4}\ mol\ dm^{-3}\ min^{-1}$, is telling us that during every minute the concentration of OH^- decreases by $6.71 \times 10^{-4}\ mol\ dm^{-3}$. Thus, over the course of four hours, it falls by 240 'lots' of $6.71 \times 10^{-4}\ mol\ dm^{-3}$ (a total of $0.161\ mol\ dm^{-3}$). A quick glance at the graph in **Figure 1.1**, however, tells us that the rate of the reaction is not constant (if it were, it would be a straight-line graph). Our value of $6.71 \times 10^{-4}\ mol\ dm^{-3}\ min^{-1}$ is the *average* rate of the reaction, taken over the whole four hours. Going back to the analogy of the two motorists covering 20 and 225 km, their respective speeds – 80 and 45 km/h, for which we could write the units as km h^{-1} – are also averaged values, taken over the whole journey: they tell us nothing about how they accelerated and decelerated over different sections of their journey.

During the first two hours of the reaction (0 to 120 min), the concentration of OH$^-$ falls from 0.200 to 0.065 mol dm^{-3} (**Figure 1.1**), from which we can calculate the average rate over this period:

$$rate = \frac{(0.200 - 0.065) \text{ mol dm}^{-3}}{(120 - 0) \text{ min}}$$

$$= \frac{0.135 \text{ mol dm}^{-3}}{120 \text{ min}}$$

$$= 1.1 \times 10^{-3} \text{ mol dm}^{-3} \text{ min}^{-1}$$

A similar calculation gives a value of 2.2×10^{-4} mol dm^{-3} min^{-1} for the rate of over the *next* two hours (120 to 240 min):

$$rate = \frac{(0.065 - 0.039) \text{ mol dm}^{-3}}{(240 - 120) \text{ min}}$$

$$= \frac{0.026 \text{ mol dm}^{-3}}{120 \text{ min}}$$

$$= 2.2 \times 10^{-4} \text{ mol dm}^{-3} \text{ min}^{-1}$$

Equipped with these values, we are now able to give a slightly more detailed description of the data presented graphically in **Figure 1.1**: the rate of reaction is 1.1×10^{-3} mol dm^{-3} min^{-1} averaged over the first two hours, but then slows down to 2.2×10^{-4} mol dm^{-3} min^{-1} over the final two hours. (The average of these two figures corresponds, allowing for the rounding error, to the rate we calculated over the full four hours, namely 6.7×10^{-4} mol dm^{-3} min^{-1}.) Notice how the slope of graph is steeper between 0 and 120 min than between 120 and 240 min. By looking at the steepness – the *gradient* – of the slope of such a plot we can judge, without the need for any calculations, how the rate of reaction changes over time: a steep slope corresponds to a high rate of reaction.

Rather than stating the average rate of a reaction over a particular time interval, it would be preferable to be able to state its rate at any given time point: what is the rate of the reaction after exactly 10 seconds, for example? Motorists, after all, are not required to work out their speed by dividing distance by time: motor cars are equipped with speedometers to report their speed continuously at every moment throughout a journey. Although there is no direct chemical equivalent to the car speedometer, what we are able to do in chemistry is calculate the rate of a reaction over ever-smaller time intervals.

When we calculate the average rate of a reaction between two time points on a concentration *vs* time graph, what we are really doing is treating the curve as through it were a straight line. **Figure 1.2** shows the same set of data as shown in **Figure 1.1**, but the right-angled triangles used to calculate the rate over three different time intervals are included. In the upper graph, used to measure the rate over the whole four hours, the slope of the curve has been taken to be that of the hypotenuse (a straight line) of a triangle whose base covers the full 240 minutes for which data has been plotted. When we divided 0.161 mol dm^{-3} by 240 min to obtain the rate of 6.71×10^{-4} mol dm^{-3} min^{-1}, what we were doing, in effect, was dividing the height of this triangle by its base. In maths, you may have been taught to calculate the gradient of straight-line graph by dividing the change on the y-axis (the 'rise' or 'fall', depending on whether the slope is positive or negative) by the change on the x-axis (the 'run'). No matter how we choose to describe this procedure, it amounts to the same thing: we are approximating the gradient of a curve to that of a straight line.

The middle and lower plots in **Figure 1.2** show the triangles used to calculate, respectively, the rate of reaction over the first 120 and 20 minutes of the reaction. Notice how the smaller the time interval over which the rate is calculated, the more closely the gradient of the curve matches or 'fits' that of the corresponding straight line. Indeed, all five of the data points 'make contact' with the hypotenuse of the triangle constructed to measure the rate over the first 20 min of the reaction (lower trace). This indicates that the rate is *approximately* constant over the first 20 min and that the gradient obtained by dividing the height of

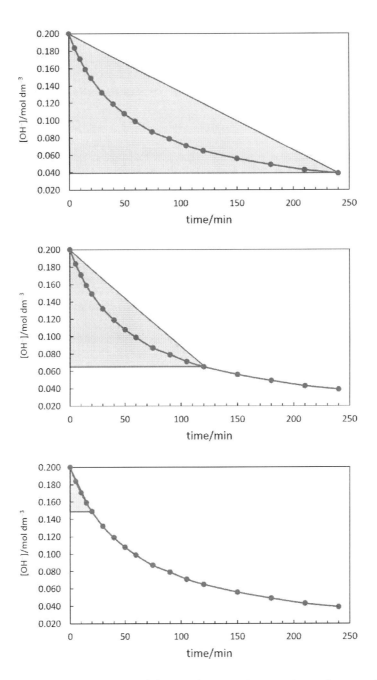

Figure 1.2   The use of right-angled triangles to estimate the rate of reaction between sodium hydroxide with bromomethane over four-hour (*upper graph*), two-hour (*middle*) and 20-minute (*lower*) time intervals. The reaction conditions are as stated in **Figure 1.1**

this triangle (0.051 mol dm^{-3}) by its base (20 min), 2.6×10^{-3} mol dm^{-3} min^{-1}, gives a good estimation of its value at any specific moment.

Although we have now obtained the rate during the first 20 minutes of the reaction, this is still an average value. We might be tempted to 'spilt the difference' and say that 2.6×10^{-3} mol dm^{-3} min^{-1} is the rate at the mid-point of the time interval over which the average was calculated, but that would be to assume that the curve is symmetrical about the 10-min point. However, **Figure 1.3**, which zooms-in on the curve over the first 20-min of the reaction, shows that this is not the case: when the hypotenuse of the triangle (which, for clarity, is shown without its other two sides) is shifted to the left, it can be seen to 'touch' the curve not at the 10-min mid-point, but just after 7 minutes.

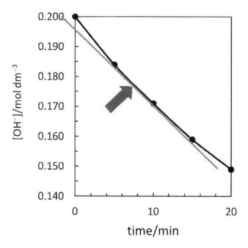

Figure 1.3   Enlarged view of the lower graph in **Figure 1.2**. The blue line is the hypotenuse of the triangle used to measure the gradient, but it has been shifted to the left until it just makes contact with the curve (indicated by the arrow).

To obtain the rate after 10 minutes, we would have to measure the gradient of the hypotenuse of a triangle that captures the slope of the curve at exactly the 10-min time-point. We could try to do this by constructing an infinitesimally small triangle, centred on the 10-min time-point, using data points from the y-axis taken immediately before and after the 10-min time-point. (The smaller the triangle, the closer the gradient of its hypotenuse would be to that of the curve at 10 minutes.)

Imagine, for example, constructing a triangle with a base extending from 9.999 to 10.001 minutes: in doing so, we could be much more confident that the gradient of the triangle's hypotenuse matches that of the curve at exactly 10 minutes. If you are studying mathematics at A-level, you will recognise this procedure as differentiation: it is based on the assertion that if you zoom-in on ever-smaller sections of any curve ('going to the limit'), it will eventually be a straight line. If you are not studying maths, then there is no need to be concerned about this. It simply helps to understand that when you measure the gradient of a curve by drawing a **tangent** to it, you are in fact constructing a tiny triangle: the gradient of the triangle's hypotenuse is the same as that of the curve at the exact point at which the two touch. The hypotenuse is then extended (and referred to as a tangent) to enable its gradient to be measured.

The gradient of a tangent drawn to a concentration *vs* time graph corresponds to the *instantaneous* rate of the reaction at the time-point at which the tangent is drawn. The rate of the reaction between sodium hydroxide and bromoethane shown in **Figure 1.1** (and **1.2** and **1.3**) is changing continuously, yet we can obtain its value at any moment by drawing a tangent. In **Figure 1.4**, the first part of the concentration *vs* time graph is shown with a tangent fitted at the 10-min point.

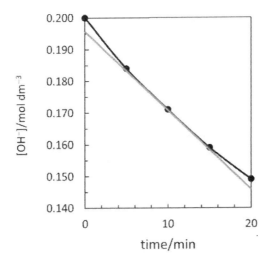

Figure 1.4   Enlarged view of the first 20 minutes of the curve shown in **Figure 1.1**. The straight line is a tangent to the curve at the 10-min time-point.

The gradient of this tangent, which captures the instantaneous rate of the reaction at exactly the 10-minute time-point, is -2.5×10^{-3} mol dm^{-3} min^{-1}. When calculating the gradient of the tangent, it is the *change* in the value on the y-axis that is divided by the time over which it has taken place. In this case, the change in the value on the y-axis is -0.0500 mol dm^{-3} (the concentration of OH^- falls from 0.196 to 0.146 mol dm^{-3} over 20 minutes). Although the gradient is negative, the minus sign is omitted when referring to the rate of reaction.

1.2 What to measure?

At the beginning of this chapter, it was stated that the reaction between bromoethane and the hydroxide ion can be monitored by measuring the change in concentration of the latter.

$$CH_3CH_2Br \quad + \quad OH^- \quad \rightarrow \quad CH_3CH_2OH \quad + \quad Br^-$$

As the equation above shows, hydroxide ions are consumed during the reaction, so a *decrease* in the concentration of OH^- (a negative change) corresponds to a *positive* rate at which bromoethane and OH^- are converted to ethanol and the bromide ion. Similarly, had we chosen to measure the rate of the reaction by monitoring the concentration of bromoethane, we would also have obtained a graph with a negative gradient. In fact, because the starting concentrations of bromoethane and sodium hydroxide were both 0.200 mol dm^{-3} (see **Figure 1.1**), and because each hydroxide ion reacts with one bromoethane molecule, a plot of CH_3CH_2Br concentration against time would look identical to the plot of OH^- concentration against time: the concentrations of bromoethane and the hydroxide ion decrease at exactly the same rate.

If, on the other hand, we had chosen to determine the rate of this reaction by measuring the change in ethanol concentration, the resultant concentration *vs* time graph would have had a *positive* gradient (**Figure 1.5**). The numerical value of the gradient taken at any chosen time point, however, would be identical to that measured from the plot of OH^- concentration against time, but its sign would be reversed (at 10 min, for example, the gradient would be $+2.5 \times 10^{-3}$ mol dm^{-3} min^{-1}). Since reactions rates are always positive, the rate would be the same, irrespective of whether we had measured the change in the

9

concentration of ethanol or the hydroxide ion against time, being 2.5×10^{-3} mol dm^{-3} min^{-1} at the 10-min time-point.

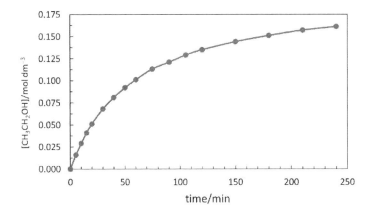

Figure 1.5 Time-course of ethanol concentration during the reaction of sodium hydroxide with bromoethane. The reaction conditions are as given above **in Figure 1.1.**

This brings us to the point where we can write a more formal definition of the rate of a reaction, which is *the rate of change in the concentration of a stated species*. This 'species' may be a reactant (such as OH$^-$) or a product (ethanol), the choice usually being based on the ease with which the various species can be measured. In the reaction between the hydroxide ion and bromoethane, the former is the obvious choice because its concentration is easy to measure by titration.

Since the balanced equation for this reaction (see above) shows that one mole of OH$^-$ reacts with one mole of CH_3CH_2Br to form one mole of CH_3CH_2OH (and one mole of Br$^-$ ions), the rate will always be the same, irrespective of which species we choose to measure. In reactions where the stoichiometry is not so simple, however, extra care must be taken to state the species (which may be a reactant or a product) whose change in concentration the rate refers to. Consider, for example, the chain-termination reaction in which two methyl radicals combine to form a molecule of ethane (a reaction you will have encountered in the study of the radical-substitution reactions between halogens and alkanes, initiated by ultraviolet light):

$$\cdot CH_3 \ + \ \cdot CH_3 \ \rightarrow \ C_2H_6$$

Since two methyl radicals 'disappear' for each ethane molecule formed, it follows that the concentration of methyl radicals will decrease at *twice* the rate at which ethane is formed. The two graphs in **Figure 1.6** show the rapid decay of methyl radicals and formation of ethane over a 1-millisecond[a] (0.001 s) time interval.[b]

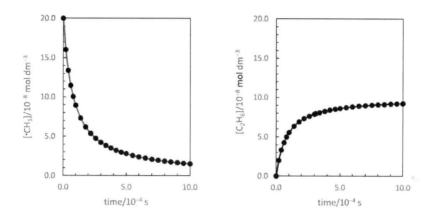

Figure 1.6   Graphs showing the rapid decrease in the concentration of methyl radicals (*left*) and increase in ethanol (*right*) concentration over a 1-ms time-period.

The curve for the decay of the methyl radical is at all points steeper than that for the formation of ethane. This is seen more clearly in **Figure 1.7**, which shows the same curves with tangents drawn at exactly 0.2 milliseconds (a randomly chosen time point). The gradient of the curve for the decay of methyl radicals (on the left) is calculated by constructing a right-angled triangle in which the tangent forms the hypothenuse: dividing the change in radical concentration by the time over which it has taken place gives the gradient – and therefore the reaction rate – at 0.2 ms. The size of the triangle used to calculate such a gradient is irrelevant because its height and base will always change together in the same ratio (doubling the height doubles the base *etc*). Defining the size of the triangle by the points at which the tangent intercepts the two axes, however, will always give the most accurate result because it means only one value is being read off each axis. In this example, we see that the

---

[a] Just as there are 1000 mm in 1 m, there are 1000 milliseconds in one second: 1 ms = 1×10^{-3} s (*i.e.* 0.001 s).
[b] Methyl radicals are generated from propanone using an extremely short pulse of ultraviolet radiation from a laser ($CH_3COCH_3 \rightarrow 2\ {}^{\bullet}CH_3 + CO$).

concentration of methyl radicals has changed by -9.8×10^{-8} mol dm^{-3} over 4.8×10^{-4} s (the height and base, respectively, of the right-angled triangle of which the tangent forms the hypotenuse). This gives a gradient of -2.0×10^{-4} mol dm^{-3} s^{-1}:

$$\text{gradient} = \frac{-9.8 \times 10^{-8} \text{ mol dm}^{-3}}{4.8 \times 10^{-4} \text{ s}}$$

$$= -2.0 \times 10^{-4} \text{ mol dm}^{-3} \text{ s}^{-1}$$

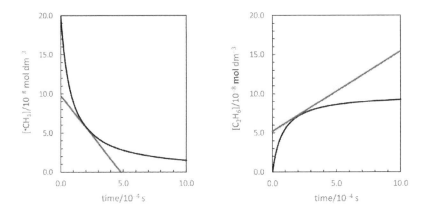

Figure 1.7   Graphs showing the decrease in the concentration of methyl radicals (*left*) and increase in ethanol (*right*) concentration over 1 ms (as shown in **Figure 1.6**, but with the omission of the data points for clarity). The line on each curve is a tangent drawn at 2.0×10^{-4} s (0.2 ms).

The gradient is negative because the value on the *y*-axis, the concentration of methyl radicals, has *decreased* with time. A similar calculation gives the gradient of the curve for the *formation* of ethane with time (**Figure 1.7**, right-hand graph): here the height of the triangle (the concentration of ethane) has changed by $+10.2 \times 10^{-8}$ mol dm^{-3} (from

12

5.1×10^{-8} to 15.3×10^{-8} mol dm^{-3})^c over 10.0×10^{-4} s, giving a gradient of $+1.0 \times 10^{-4}$ mol dm^{-3} s^{-1}. Both of these gradients correspond to the rate of the reaction at precisely 0.2 ms, but as the methyl radicals are decaying at the twice the rate at which ethane is being formed, it is essential we state the species to which we are referring: in terms of the change in concentration of methyl radicals, the rate is 2.0×10^{-4} mol dm^{-3} s^{-1} (with the minus sign dropped); but in terms of the formation of ethane, the rate is 1.0×10^{-4} mol dm^{-3} s^{-1}.

1.3   Getting physical – the need for collisions

For any chemical reaction to take place, collisions must occur between the reactants: if the reactants do not collide, no reaction can take place. Increasing the reactant concentration results in an increase in the number of collisions taking place each second (the so-called *collision frequency*); however, not all collisions result in a reaction: if the reactants do not collide with sufficient energy, or collide in the 'wrong' orientation, they will simply bounce off each other unchanged. This model of reaction rates, which you will no doubt remember from GCSE chemistry, is known as **collision theory**. We will look in detail at how the energy of these collisions (and their orientation) affects the reaction rate in Chapter 2, but first we will focus our attention on the effects of reactant concentration on reaction rates.

We saw above how the rate of the reaction between the hydroxide ion and bromoethane decreases with time. We might attribute this to the fact that the concentrations of both reactants decrease over time (they are being converted into products), so there are progressively fewer and fewer OH$^-$ ions and CH_3CH_2Br molecules to collide with each other as the reaction runs its course. During the reaction, the carbon-to-bromine bond in CH_3CH_2Br breaks and a new bond is formed between the

---

^{c} It is acknowledged that the values 10.2×10^{-8} and 15.3×10^{-8} are not in the correct format (in standard form, they should be 1.02×10^{-7} and 1.53×10^{-7}). In chemistry, you will often encounter data sets with such a large range they span more than one 'order of magnitude' (factor of 10), thereby forcing us to put some of the numbers in standard form to an 'inappropriate' power of 10 (*e.g.* the maximum value on the *y*-axis in **Figure 1.7** is given as 20.0×10^{-8} rather than 2.00×10^{-7}). To minimise the risk of mistakes during calculations, it is perfectly acceptable to leave such numbers in the inappropriate standard form *during* calculations, but the *final* answer should be given in correct standard form.

oxygen atom in the hydroxide ion and the carbon that was bonded to bromine. In A-level textbooks, the reaction mechanism is typically written as follows, in which bond breaking and formation occur in a direct collision between a CH_3CH_2Br molecule and an OH^- ion:[d]

It follows, therefore, that anything we can do to increase the frequency of collisions between CH_3CH_2Br and OH^- will increase the rate of this reaction. In other words, increasing the concentrations of CH_3CH_2Br and OH^-, individually or together, will increase the rate. Thus, doubling the concentration of CH_3CH_2Br will double the rate, whereas doubling the concentration of CH_3CH_2Br *and* OH^- will result in a 4-fold increase in the rate: doubling [CH_3CH_2Br][e] doubles the rate, but then doubling [OH^-] doubles it again. No surprises there, then – everything so far is consistent with the collision theory covered at GCSE level.

Now consider the nucleophilic substitution reaction that takes place when a cold, dilute, aqueous solution of sodium hydroxide solution[f] is added to 2-bromomethylpropane, which is a *tertiary* halogenoalkane, forming methylpropan-2-ol:[g]

---

[d] Notice how the reaction results in an inversion of the configuration of the molecule around the carbon atom where the reaction takes place – in much the same way that an umbrella might invert when caught by a strong wind.

[e] Square brackets are used to represent the concentration of a species. Thus [CH_3CH_2Br] means 'concentration of CH_3CH_2Br'.

[f] Halogenoalkanes (also called haloalkanes), particularly tertiary halogeno-alkanes, also undergo elimination reactions with hydroxide ions. To favour substitution over elimination, the alkali is added as a cold, dilute, aqueous solution. The elimination reaction is favoured by using a hot, concentrated, ethanolic solution of the alkali (as covered, for example, in the current AQA chemistry course).

[g] This halogenoalkane and the alcohol are often named 2-bromo-2-methylpropane and 2-methylpropan-2-ol, respectively, but it is not necessary to state that the methyl group is on carbon 2.

$$(CH_3)_3CBr \; + \; OH^- \; \rightarrow \; (CH_3)_3COH \; + \; Br^-$$

You would be forgiven for expecting the rate of this reaction responds to changes in the concentrations of $(CH_3)_3CBr$ and OH^- in exactly the same manner as seen in the corresponding reaction involving CH_3CH_2Br. This, however, is not the case. Although sensitive to changes in $[CH_3CH_2Br]$, the rate is *not* affected by changes in $[OH^-]$. This somewhat counterintuitive behaviour can be explained if the reaction is envisaged to occur in two discrete steps. Initially, in the slower step, the carbon-to-bromine bond in the halogenoalkane breaks, forming a carbocation intermediate; only then does the hydroxide ion become involved, forming a covalent bond with the carbon (which now bears a full positive charge) in the faster step:

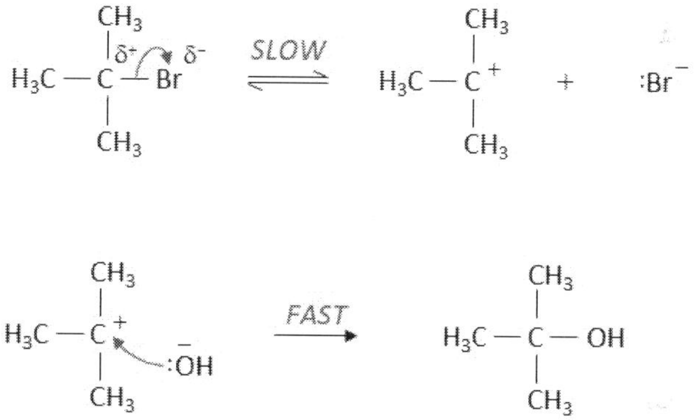

The first step, in which $(CH_3)_3CBr$ undergoes heterolytic bond fission to form $(CH_3)_3C^+$ and Br^-, is an extremely slow process compared with the subsequent, rapid addition of OH^- to the carbocation. Nevertheless, no matter how miniscule the proportion of $(CH_3)_3CBr$ molecules that are dissociated into $(CH_3)_3C^+$ and Br^- at any particular moment, increasing the concentration of $(CH_3)_3CBr$ will increase the number of $(CH_3)_3C^+$ ions available to combine with OH^- and will, therefore, increase the rate of the *overall* reaction.

Increasing the concentration of the OH^- ion can only increase the rate of the overall reaction if there is enough $(CH_3)_3C^+$ available to react with the extra OH^-. However, because the rate of $(CH_3)_3C^+$ formation is very slow, and its removal is very fast, its concentration will always be tiny in comparison with that of the hydroxide ion. In other words, $(CH_3)_3C^+$ will always be the limiting reactant in the second step. Increasing or decreasing the concentration of OH^- does not affect the overall rate because it is never in short supply: there will always be more than enough OH^- ions present to combine with the comparatively tiny amount of $(CH_3)_3C^+$.

Where a reaction takes place in two or more steps, the slowest step – which determines the rate of the overall rate of the reaction – is called the **rate-determining step** (RDS). You may also see this referred to as the rate-limiting step (RLS). When you measure the rate of such a reaction (*e.g.* the nucleophilic substitution described here) you are, in fact, measuring the rate of only the RDS. That is why changing the concentration of any reactant that is not involved in the RDS does not affect the rate you are measuring. One can think of numerous analogies to the RDS from everyday life. Imagine, for example, you are making a cake. After baking it in the oven for 20 minutes (the RDS), a cherry is placed on the top, which takes 1 second. Trying to speed up the production process by standing by with hundreds of cherries, ready to place one on the cake as soon as it is removed from of the oven would, for obvious reasons, be futile. In this analogy, the cherries are equivalent to the hydroxide ions in our nucleophilic substitution example. In the coming chapters, we will look at the kinetics of several other multi-step reactions.

When they are first introduced to organic reaction mechanisms, students often ask how chemists know that reactions take place through the various sequences of individual steps they are being taught. How do chemists know, for example, that, whereas CH_3CH_2Br and OH^- react in a single step, the corresponding reaction involving $(CH_3)_3CBr$ proceeds through a two-step mechanism? The answer to this question is, of course, through kinetics: the initial aim of any investigation into the kinetics of a particular reaction is to determine how changing the concentration of each of the reactants affects its rate. This involves measuring the rate of the reaction in a series of experiments in which the concentration of each reactant is varied in turn: by keeping the concentrations of all the other reactants constant, we are 'isolating' the effect of changing the concentration of the selected reactant. Having established how changing the concentration of each reactant affects the rate of the overall reaction, the chemist must then propose a mechanism that is consistent with the observed behaviour. Only reactants whose concentration affect the overall rate are included in the rate-determining set.

From a knowledge of the chemical properties of the reactants, it is often possible to make an 'educated guess' as to how a reaction might take place. For example, you may have been taught in Year 12 that tertiary carbocations, such as $(CH_3)_3C^+$, are *relatively* stable compared with their primary cousins, which are generally considered to be too unstable to be formed as reaction intermediates. It is therefore reasonable to expect that, whereas CH_3CH_2Br and OH^- react in a single step, without the formation of a carbocation, the corresponding reaction involving $(CH_3)_3CBr$ proceeds *via* the two-step mechanism described above. Any proposed mechanism, however, *must* be confirmed experimentally. Kinetics do not necessarily prove that a proposed reaction mechanism is valid, but they can certainly eliminate an incorrect mechanism. Having proposed a mechanism that is consistent with the kinetics, a chemist might then seek other, independent lines of supporting evidence. Let's now look at how kinetics supports the single-step mechanism for the reaction between CH_3CH_2Br and OH^-.

A common technique used to measure the rate of a reaction at A-level is the so-called the **initial-rates method**, in which the rate is measured at the instant the reaction is started. This entails drawing a tangent at the zero time-point of a concentration *vs* time graph. **Figure 1.8** shows the results of such an investigation into effect of changing the concentration of bromoethane on the rate of its reaction with the hydroxide ion (from sodium hydroxide). The initial concentration of CH_3CH_2Br was increased from 0.050 mol dm^{-3} in **(a)** up to 0.400 mol dm^{-3} in **(f)**. As the same initial concentration of NaOH (0.200 mol dm^{-3}) was used in all six experiments, it follows that the increase in the reaction rate seen in going from **(a)** to **(f)** must be due to the increase in the concentration of CH_3CH_2Br and that alone: the experiments have allowed us to observe *in isolation* the effect of changing the CH_3CH_2Br concentration on the rate of the reaction.

The next step is to determine exactly how sensitive the rate of the reaction is to changes in the concentration of the reactant whose concentration is being changed: does, for example, a two-fold increase in the concentration of CH_3CH_2Br cause the rate to increase two-fold, or does the rate increase by a greater factor? This necessitates measuring the gradient of the tangent drawn to each curve at its zero time-point,

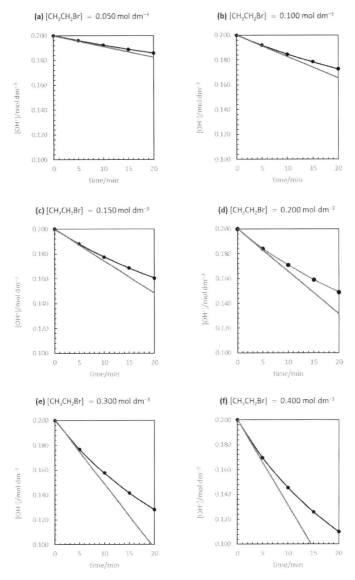

Figure 1.8 A series of six experiments to investigate how varying the initial concentration of bromoethane affects the rate of its reaction with the hydroxide ion at 50 °C. The initial concentration of bromoethane was increased from 0.05 to 0.400 mol dm⁻³, as indicated in graphs **(a)** to **(f)**. In all experiments, the initial concentration of sodium hydroxide was 0.200 mol dm⁻³. The straight lines are tangents drawn to each curve at the zero time-point.

which gives the initial rate of the reaction for each initial concentration of CH₃CH₂Br. Taking curve **(f)** as an example, we see that over a period of 876 seconds (14.6 min) the concentration of OH⁻ ions fell by

0.100 mol dm^{-3} (from an initial concentration of 0.200 mol dm^{-3} to 0.100 mol dm^{-3}), which gives a value of 1.14 mol dm^{-3} s^{-1} for the initial rate of reaction when the concentration of CH$_3$CH$_2$Br is 0.400 mol dm^{-3}. The initial rates calculated from the tangents shown in all six experiments are given in **Table 1.1**.

[CH$_3$CH$_2$Br]/mol dm^{-3}	rate/mol dm^{-3} s^{-1}
0.050	1.43 × 10^{-5}
0.100	2.86 × 10^{-5}
0.150	4.29 × 10^{-5}
0.200	5.72 × 10^{-5}
0.300	8.58 × 10^{-5}
0.400	1.14 × 10^{-4}

Table 1.1   Effect of CH$_3$CH$_2$Br concentration on the rate of its reaction with the OH$^-$ ion (which was present at a concentration of 0.200 mol dm^{-3} in all experiments). The rates were calculated using the initial rates method, from the tangents to the plots shown in **Figure 1.8**.

The collected data show that, for this reaction, increasing the concentration of CH$_3$CH$_2$Br by a given factor causes the initial rate to increase by the *same* factor. For example, doubling [CH$_3$CH$_2$Br] from 0.050 to 0.100 mol dm^{-3} causes the rate to double, whereas increasing [CH$_3$CH$_2$Br] 8-fold, to 0.400 mol dm^{-3}, causes a corresponding 8-fold in the rate. Similar experiments, in which [CH$_3$CH$_2$Br] was held constant and [OH$^-$] was varied, have established that the reaction rate responds to changes in OH$^-$ concentration in exactly the same manner.

You may be wondering what the point of all this is. Are we not simply stating the obvious? Could these results have not been anticipated through reasoning and common sense? As already stated above, however, such relationships must be established experimentally: reaction mechanisms are proposed on the basis of the observed kinetics, not the other way around. Thus, in contrast to its reaction with CH$_3$CH$_2$Br, we saw how changing the concentration of OH$^-$ does not affect the rate of its reaction with $(CH_3)_3CBr$. Later in this chapter (and in others), we will examine reactions in which doubling the concentration of a reactant causes the rate to more than double, but in the first instance the aim is to introduce the underlying principles of kinetics using very simple examples. If, therefore, you are finding the above reasoning to be so simple and obvious that you are questioning why it needs to be stated at all, then this is succeeding

Plotting the data from **Table 1.1**, $[CH_3CH_2Br]$ against initial rate, gives a straight line (**Figure 1.9**), showing that the rate of reaction is proportional to the concentration of the halogenoalkane:

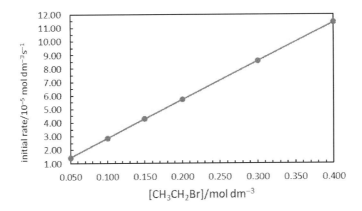

Figure 1.9   Effect of CH_3CH_2Br concentration on the rate of its reaction with the OH^- ion (which was present at a concentration of 0.200 mol dm^{-3} in all experiments). The data points are from **Table 1.1**.

The equation of a straight line has the form $y = mx + c$, where m is the gradient (a constant) and c is a constant that can be determined from the intercept on the y-axis (the value of y when x is zero). As the straight-line plotted in **Figure 1.9** passes through the origin[h], c is zero, leaving us with

---

[h] Although the data point has not been plotted in **Figure 1.9**, the initial rate must be zero when there is no CH_3CH_2Br present.

the equation $y = mx$. Since y represents the (initial) rate and x the concentration of CH_3CH_2Br, we can write:

$$\text{rate} = m \times [CH_3CH_2Br]$$

We can obtain m by measuring the gradient of the straight line directly from the graph: increasing $[CH_3CH_2Br]$ by 0.35 mol dm^{-3} (from 0.050 to 0.400 mol dm^{-3}) causes the initial rate to increase by 9.97×10^{-5} mol dm^{-3} s^{-1} (from 1.43×10^{-5} to 1.14×10^{-4} mol dm^{-3} s^{-1}):

$$m = \frac{9.97 \times 10^{-5} \text{ mol dm}^{-3} \text{ s}^{-1}}{0.35 \text{ mol dm}^{-3}}$$

$$= 2.85 \times 10^{-4} \text{ s}^{-1}$$

Notice, as previously, how the units of the gradient have been derived by cancelling out those of the two variables used in its calculation. We can now write a 'formula' (an expression) to 'fit' the data given in **Figure 1.9** (and **Table 1.1**):

$$\text{Rate} = (2.85 \times 10^{-4} \text{ s}^{-1}) \times [CH_3CH_2Br]$$

To obtain the rate at any given concentration of bromoethane, then, we simply multiply $[CH_3CH_2Br]$ by the constant 2.85×10^{-4} s^{-1}. There is, however, one major drawback with this handy little expression: it works only when the concentration of OH$^-$ is 0.200 mol dm^{-3} and, as we saw above, the rate of this reaction is affected by the concentration of CH_3CH_2Br *and* that of OH$^-$.

In the series of experiments shown in **Figure 1.8**, the concentration of OH$^-$ was held at a constant 0.200 mol dm^{-3}, which prevented it from affecting the rate. Although no data was shown, it was described earlier how changes in the concentration of OH$^-$ affect the rate in exactly the same manner as changes in the concentration of CH_3CH_2Br. Thus, doubling [OH$^-$] doubles the rate, whereas increasing [OH$^-$] 4-fold causes a corresponding 4-fold in the rate *etc*. This is true whatever the concentration of CH_3CH_2Br happens to be.

If, instead of keeping the concentration of OH$^-$ at a constant 0.200 mol dm^{-3} in the experiments described in **Figure 1.8**, we had kept it at, for example, 0.400 mol dm^{-3}, we would have found the initial rate at every concentration of CH_3CH_2Br tested to be double the value shown in **Table 1.1**. Similarly, the slope of the straight line plotted in **Figure 1.9** would have been twice as steep; in other words, m would have doubled to $5.70 \times 10^{-4} s^{-1}$. Thus, m is a constant, with a value of $2.85 \times 10^{-4} s^{-1}$, only when the concentration of OH^- is 0.200 mol dm^{-3}. To reflect the fact that the reaction rate depends on the concentration of both CH_3CH_2Br and OH^-, we can modify the expression,

$$rate = m \times [CH_3CH_2Br]$$

to the more general expression,

$$rate = k \times [OH^-] \times [CH_3CH_2Br]$$

where k is a yet-to-be-determined constant that enables us to calculate the rate when we know the concentration of CH_3CH_2Br at any concentration of OH^-.

In the series of experiments in which the concentration of OH^- was kept at a fixed value of 0.200 mol dm^{-3}(**Figure 1.8**), this general expression contained two constants, namely k and $[OH^-]$:

$$rate = k \times 0.200 \text{ mol dm}^{-3} \times [CH_3CH_2Br]$$

Multiplying together two constants always gives another constant. It follows, then, that m in our original expression must represent k multiplied by 0.200 mol dm^{-3},

$$rate = k \times \underbrace{0.200 \text{ mol dm}^{-3} \times}_{m} [CH_3CH_2Br]$$

that is to say,

$$m = k \times 0.200 \text{ mol dm}^{-3}$$

from which k can be calculated by rearranging and substituting in the value of m obtained from the gradient of the plot in **Figure 1.9**:

$$k = \frac{2.85 \times 10^{-4} s^{-1}}{0.200 \text{ mol dm}^{-3}}$$

$$= 1.43 \times 10^{-3} \text{ dm}^3 \text{ mol}^{-1} s^{-1}$$

We now have a formula that enables us to calculate the initial rate of the reaction for any given concentrations of OH^- and CH_3CH_2Br,

$$\text{rate} = k \times [OH^-] \times [CH_3CH_2Br]$$

which we can back-test using any pair of values from **Table 1.1**. For example, with $[CH_3CH_2Br]$ equal to 0.400 mol dm^{-3} and $[OH^-]$ at 0.200 mol dm^{-3} we have:

$$\text{rate} = (1.43 \times 10^{-3}\,\text{dm}^3\,\text{mol}^{-1}\,\text{s}^{-1})(0.200\,\text{mol dm}^{-3})(0.400\,\text{mol dm}^{-3})$$

$$= 1.14 \times 10^{-4}\,\text{mol dm}^{-3}\,\text{s}^{-1}$$

Notice, again, how the units cancel down nicely to leave the appropriate units for rate (a dm^3 and dm^{-3} pair cancel out, as do mol and mol^{-1}). Using this expression, which is called a **rate equation**, it is easily demonstrated that doubling the concentration of OH^- while keeping $[CH_3CH_2Br]$ the same causes the rate to double:

$$\text{rate} = (1.43 \times 10^{-3}\,\text{dm}^3\,\text{mol}^{-1}\,\text{s}^{-1})(0.400\,\text{mol dm}^{-3})(0.400\,\text{mol dm}^{-3})$$

$$= 2.29 \times 10^{-4}\,\text{mol dm}^{-3}\,\text{s}^{-1}$$

The constant k is the is called the **rate constant** of the reaction. A rate constant is specific to a particular reaction. As we shall see in the next chapter, rate constants are temperature dependent: they apply only for a stated temperature. (We will also learn in the next chapter that this is because the value of a rate constant depends primarily on the activation energy of the reaction.) If you look back at the legend for **Figure 1.1**, you will see that the data from which we calculated the rate constant for the reaction between OH^- and CH_3CH_2Br was obtained at $50\,°C$, so it is only at this temperature that k has a value of 1.43×10^{-3} dm^3 mol^{-1} s^{-1}.

Although the rate of a given reaction is affected by changes in reactant concentration (in accordance with the rate equation), rate constants themselves are *not* affected by reactant concentrations. Other than temperature, the only experimental variable that can change the value of a rate constant is the nature of the solvent in which the reaction is carried out; such effects, however, are not covered at A-level. Although the rate constant of a reaction speeded up by a catalyst is greater than that of the uncatalysed reaction, here we are comparing – effectively – two different reactions: catalysts change the 'route' by which reactants are converted to products (the reaction mechanism changes), so although the reactants and products are the same as in the uncatalysed reaction, the activation energy is different, which in turn changes the rate constant. These points will become much clearer in the next chapter, but it is important they are made at the outset.

Having introduced the concepts of the rate equation and the rate constant using a relatively straightforward example, we will now look in some detail at a reaction displaying more complicated kinetics.

1.4 Oxidation of bromide ions by bromate(V) in acidic solution

The overall equation for the reaction that occurs between bromate(V) and bromide ions under acidic conditions is:

$$BrO_3^- \ + \ 5\,Br^- \ + \ 6\,H^+ \ \rightarrow \ 3\,Br_2 \ + \ 3\,H_2O$$

It would be quite possible to analyse the kinetics of this reaction without much of an appreciation of the underlying chemistry, treating it purely as a number-crunching exercise. You should, however, always try to gain a 'feel' for what is going on in a reaction – try to get 'under its skin', so to speak. Approaching problems from a chemical – rather than a purely mathematical – perspective will develop your confidence and skills as a chemist. At the end of the day, you are not going to be awarded an A-level in chemistry just for doing a bit of relatively trivial maths: the examination questions are designed to assess your abilities as a chemist. When you are confronted with an unfamiliar reaction (other than perhaps in organic chemistry), it is a good idea to first ask yourself whether it is a redox reaction or an acid-base reaction. It is then usually relatively straightforward to get to the heart of the reaction and see what is going on.

In this reaction, the oxidation state of bromine is changing, so it is clearly a redox reaction. One bromine atom is being reduced, from +5 in BrO_3^- to zero in Br_2, and five are being oxidised, from −1 in Br^- to zero in Br_2. Having used oxidation numbers to spot these changes, we can then write down the two half-equations which, when added together and after cancelling out the electrons, give the overall equation for the reaction:[i]

$$2 \, BrO_3^- \ + \ 10 \, e^- \ + \ 12 \, H^+ \ \rightarrow \ Br_2 \ + \ 6 \, H_2O$$

$$10 \, Br^- \ \rightarrow \ 5 \, Br_2 \ + \ 10 \, e^-$$

$$\overline{2 \, BrO_3^- \ + \ 10 \, Br^- \ + \ 12 \, H^+ \ \rightarrow \ 6 \, Br_2 \ + \ 6 \, H_2O}$$

It is now easier to see what is going on: the 10 electrons required to reduce the two BrO_3^- ions to Br_2 are obtained from the ten Br^- ions that are being oxidised to form five Br_2 molecules. Cancelling down (and dividing through by 2) gives us the original equation:

$$BrO_3^- \ + \ 5 \, Br^- \ + \ 6 \, H^+ \ \rightarrow \ 3 \, Br_2 \ + \ 3 \, H_2O$$

Although the overall equation for the reactions shows that one BrO_3^- ion reacts with 5 Br^- ions and 6 H^+ ions, this could never take place in a simultaneous collision between all 12 particles. We would, therefore, expect the reaction to take place in a series of steps. The slowest step – the rate-determining step – controls the rate of the overall reaction. Recall that when we measure the rate of a multi-step reaction, we are in fact measuring only the rate of the RDS, therefore only the reactants involved in this step will appear in the rate equation (in the ratio – the stoichiometry – in which they are involved in the RDS).

Table 1.2 gives a summary of the data from a series of experiments investigating the effect of changing the reactant concentrations on the rate of this reaction. This table 'format' is typical of what you will see in the examination questions. Using this data, you may be asked to derive the rate equation and the rate constant for the reaction, which will be undertaken here. Before we begin this task, however, we will first consider what exactly is meant by the 'rate' of this reaction, which is

---

[i] See Appendix I for a description of the procedure for constructing and combining half-equations.

Experiment	$[BrO_3^-]$ /mol dm^{-3}	$[Br^-]$ /mol dm^{-3}	$[H^+]$ /mol dm^{-3}	initial rate/ mol dm^{-3} s^{-1}
1	1.50×10^{-3}	1.00×10^{-1}	5.00×10^{-2}	7.20×10^{-7}
2	3.00×10^{-3}	1.00×10^{-1}	5.00×10^{-2}	1.44×10^{-6}
3	1.50×10^{-3}	2.00×10^{-1}	2.00×10^{-1}	2.30×10^{-5}
4	1.50×10^{-3}	3.00×10^{-1}	5.00×10^{-2}	2.16×10^{-6}

Table 1.2   Effect of reactant concentrations on the initial rate of bromide ion oxidation by the bromate(V) ion under acidic conditions (at 25 °C). See text for details.

usually taken to be the change in the concentration of BrO_3^- per unit time:

$$\text{rate} = \frac{\text{change in } [BrO_3^-]}{\text{time}}$$

Despite the rate being *expressed* in terms of the change in concentration of BrO_3^-, it is usually *measured* by monitoring the formation of Br_2, which imparts a characteristic orange-brown colour to the solution ('bromine water'). Research chemists measure Br_2 concentrations using a colorimeter (spectrophotometer).[i] Since *three* molecules of Br_2 are produced for each BrO_3^- ion that reacts, the rate of Br_2 formation is divided by three to obtain the rate of BrO_3^- loss.

The first thing you should do when confronted with a collection of data on rates, such as that in given in **Table 1.2**, is find a pair of experiments in which the concentration of only one reactant has been changed. Comparing Experiments 1 and 2, for example, we see that the concentration of the BrO_3^- ion has been doubled, but the concentrations of the Br^- and H^+ ions have not been changed. We can conclude, therefore, that the doubling in rate seen between Experiments 1 and 2 (from 7.20×10^{-7} to 1.44×10^{-6} mol dm^{-3} s^{-1}) must have been

---

[i] In schools, the formation of Br_2 is often measured in a so-called clock reaction (described in Chapter 4). See, for example, Core Practical 14 in the Edexcel A-level Chemistry course.

caused by the doubling in BrO_3^- concentration: Experiments 1 and 2 have enabled us to observe *in isolation* the effect of $[BrO_3^-]$ on the rate.

Similarly, by comparing Experiments 1 and 4 we can see how the rate responds to changes in the concentration of Br^-. Increasing $[Br^-]$ by a factor of 3 has caused the rate to increase by the same factor (from 7.20 $\times$ 10^{-7} to 2.16 $\times$ 10^{-6} to mol dm^{-3} s^{-1}). Unlike the situation with the BrO_3^- and Br^- ions, there is not a pair of experiments in which the concentration of only the H^+ ion has been changed; however, equipped with our knowledge of how $[BrO_3^-]$ and $[Br^-]$ affect the rate, it is now possible to deduce the effect of changing $[H^+]$. In Experiment 3, for example, the concentration of Br^- is double that in Experiment 1: we know that this *alone* will cause the rate to double, from 7.20 $\times$ 10^{-7} to 1.44 $\times$ 10^{-6} to mol dm^{-3} s^{-1}. The actual rate in Experiment 3, however, is 2.30 $\times$ 10^{-5} mol dm^{-3} s^{-1}, which means that the 4-fold increase in $[H^+]$ has caused a *further* 16-fold increase in the rate, from 1.44 $\times$ 10^{-6} up to 2.30 $\times$ 10^{-5} to mol dm^{-3} s^{-1}. Because the rate has increased by the square of the increase in H^+ concentration ($16 = 4^2$), the reaction is said to be **second order with respect to H^+.**

The simplest way to determine the **order** of a reaction, with respect to a particular reactant, is to express the effect of a concentration change on the rate in the form:

> **An *x*-fold change in the concentration of the reactant causes an x^n-fold change in the rate, where n is the order of the reaction with respect to the reactant.**

In our example, a 4-fold increase in $[H^+]$ causes a 4^2-fold increase in the rate, so the reaction is said to be **2nd order** with respect to H^+. Similarly, the reaction is said to be **1st** order with respect to BrO_3^- because a 2-fold increase $[BrO_3^-]$ caused a 2^1-fold increase in the rate (Experiments 1 and 2). As a 3-fold increase $[Br^-]$ caused a 3^1-fold increase in the rate (Experiments 1 and 4), the reaction is also **1st** order with respect to Br^-.

The *overall* order of this reaction is 4, which is simply the sum of the orders with respect to the individual reactants $(2 + 1 + 1)$.

It is very important, when giving the order of a reaction, to indicate clearly whether this is the overall order or the order with respect to an individual reactant: unless stated otherwise, when the order of a reaction is given this is taken to be the overall order. The phrase 'with respect to' is typically shortened to 'w.r.t.', but if you use this abbreviation in the examinations you should write it in full followed by (w.r.t.) at the first usage. Instead of w.r.t., you may see 'in'. For example, the reaction between BrO_3^- and Br^- ions in the presence of acid is said to be first order *in* BrO_3^-, first order *in* Br^- and second order *in* H^+.

Once the order of a reaction with respect to each reactant has been determined, it is a simple matter to write the rate equation, in which the power to which the concentration of each reactant is raised is its order:

$$\text{rate} = k \times [BrO_3^-]^1 \times [Br^-]^1 \times [H^+]^2$$

which is simply,

$$\text{rate} = k[BrO_3^-][Br^-][H^+]^2$$

The value of the rate constant is obtained by rearranging the rate equation,

$$k = \frac{\text{rate}}{[BrO_3^-][Br^-][H^+]^2}$$

and substituting in a set of data from **Table 1.2**. Using the data from Experiment 1, for example, gives:

$$k = $$

$$\frac{7.20 \times 10^{-7} \text{ mol dm}^{-3} \text{ s}^{-1}}{(1.50 \times 10^{-3} \text{ mol dm}^{-3})(1.00 \times 10^{-1} \text{ mol dm}^{-3})(5.00 \times 10^{-2} \text{ mol dm}^{-3})^2}$$

Notice how the units of the individual values from Experiment 1 have also been entered in the rearranged equation. This will enable the units of the rate constant to be derived. If you are studying mathematics at A-level, you will have no difficulties cancelling down the units to obtain those of k. For students less confident in maths, I often suggest they 'expand' any terms raised to a power by writing them in the form of a product. The term $(5.00 \times 10^{-5} \text{ mol dm}^{-3})^2$ here can be entered as the product $(5.00 \times 10^{-5} \text{ to mol dm}^{-3}) \times (5.00 \times 10^{-5} \text{ to mol dm}^{-3})$. This avoids the need to square the units (which would become $\text{mol}^2 \text{ dm}^{-6}$), thereby simplifying the cancelling down process,

$$k \quad = $$

$$\frac{7.20 \times 10^{-7} \;\cancel{\text{mol}}\; \cancel{\text{dm}^{-3}}\; \text{s}^{-1}}{(1.50 \times 10^{-3} \;\cancel{\text{mol}}\; \cancel{\text{dm}^{-3}})\, (1.00 \times 10^{-1} \text{ mol dm}^{-3})(5.00 \times 10^{-2} \text{ mol dm}^{-3})^2}$$

giving:

$$k \quad = \quad \frac{1.92 \; \text{s}^{-1}}{(\text{mol dm}^{-3})(\text{mol dm}^{-3})(\text{mol dm}^{-3})}$$

Units are squared and cubed *etc* in exactly the same manner as are numbers, by *adding* their powers; therefore $\text{mol} \times \text{mol} \times \text{mol}$, which is really $\text{mol}^1 \times \text{mol}^1 \times \text{mol}^1$, equals mol^3. Similarly, $\text{dm}^{-3} \times \text{dm}^{-3} \times \text{dm}^{-3}$ equals dm^{-9}, so now have:

$$k \quad = \quad \frac{1.92 \; \text{s}^{-1}}{\text{mol}^3 \, \text{dm}^{-9}}$$

As we saw at the beginning of the chapter (page 2), units can be gathered together on a single line by reversing the sign of their powers when they are moved from the denominator to the numerator, giving:

$$k \quad = \quad 1.92 \; \text{dm}^9 \, \text{mol}^{-3} \, \text{s}^{-1}$$

The convention when writing units is to put those with positive powers (dm^9) before those with negative powers; they are then placed in alphabetical order (mol^{-3} coming before s^{-1}).

It is evident from the two examples we have covered so far that the units of a given rate constant depend on the overall order of the reaction. The rate constants of fourth-order reactions have units of $dm^9\ mol^{-3}\ s^{-1}$, whereas those of second-order reactions have units of $dm^3\ mol^{-1}\ s^{-1}$ (*e.g.* for the reaction between CH_3CH_2Br and OH^-). In due course, you will see that the rate constants of first- and third-order reactions have units of s^{-1} and $dm^6\ mol^{-2}\ s^{-1}$, respectively. Do not, however, attempt to memorise these sets of units; instead, get used to deriving units from your calculations. (Save your memory space for the topics where you simply have to memorise facts, such as the colours of all those transition metal-ion complexes!) This can be performed as a separate little calculation 'on the side'. For example, after working out the numerical value of the rate constant for the reaction between BrO_3^- with Br^- ions in the presence of acid (1.92), the appropriate units are given by:

$$k\ =\ \frac{\cancel{mol}\ \cancel{dm^{-3}}\ s^{-1}}{(\cancel{mol}\ \cancel{dm^{-3}})\ (mol\ dm^{-3})(mol\ dm^{-3})(mol\ dm^{-3})}$$

$$=\ \frac{s^{-1}}{mol^3\ dm^{-9}}$$

$$=\ dm^9\ mol^{-3}\ s^{-1}$$

Once you have derived a rate equation and its rate constant, it is a good idea to 'test' them out on another set of data. The data in **Table 1.2** tells us, for example, that with the reactant concentrations used in Experiment 4, the rate of the reaction is 2.16×10^{-6} to $mol\ dm^{-3}\ s^{-1}$, which it should be possible to confirm:

$$rate\ =\ k[BrO_3^-][Br^-][H^+]^2$$

$$=\ 1.92\ \times (1.50 \times 10^{-3}) \times (3.00 \times 10^{-1}) \times (5.00 \times 10^{-2})^2$$

$$=\ 2.16 \times 10^{-6}$$

By checking to see that the appropriate units of rate 'fall out' of this calculation,[k] we can confirm that we have derived the correct units for k:

$$\text{units of rate} = (\text{dm}^9\,\text{mol}^{-3}\,\text{s}^{-1})(\text{mol dm}^{-3})(\text{mol dm}^{-3})(\text{mol dm}^{-3})^2$$

$$= (\text{dm}^9\,\text{mol}^{-3}\,\text{s}^{-1})(\text{mol dm}^{-3})(\text{mol dm}^{-3})(\text{mol dm}^{-3})(\text{mol dm}^{-3})$$

$$= (\text{dm}^9\,\text{mol}^{-3}\,\text{s}^{-1})(\text{mol}^4\,\text{dm}^{-12})$$

$$= \text{mol dm}^{-3}\,\text{s}^{-1}$$

As well as enabling us to calculate the rate of a reaction under any given set of reactant concentrations, we saw earlier how the rate equation reflects the mechanism of the reaction. The overall equation for the oxidation of bromide ions by bromate(V) under acidic conditions (repeated below for convenience) indicates that the five electrons made available when five Br^- ions each lose one electron are transferred to a single BrO_3^- ion, which is thereby reduced to the element (from oxidation state +5 to zero).

$$BrO_3^- + 5\,Br^- + 6\,H^+ \rightarrow 3\,Br_2 + 3\,H_2O$$

The rate equation for this reaction (also repeated below) tells us that the overall rate of the reaction is controlled by a step involving one BrO_3^-, one Br^- and two H^+ ions.

$$\text{rate} = k[BrO_3^-][Br^-][H^+]^2$$

You would be very unlucky indeed to be asked in an A-level examination to deduce the mechanism of this reaction based solely on the rate equation. With such a complicated reaction, you are more likely to be given some form of framework within which to work; for example, you may be given a series of reaction steps and asked to suggest which one is the rate-determining step. Having said that, the questions under the new, post-2015 specifications can be very challenging indeed and are designed to weed out candidates working under the misapprehension

---

[k] The powers are being added. Thus, in the third line below, mol^{-3} plus mol^4 gives mol; and dm^9 plus dm^{-12} gives dm^{-3}.

that reaction kinetics is nothing more than GCSE algebra with x and y replaced by chemical symbols. The most challenging questions rarely involve only a single subject area; instead, they require candidates to draw upon their wider subject knowledge, applying the principles taught in a particular area to problems in a different context. In the reaction being considered here, for example, we have seen already how the application of redox chemistry (oxidation states, half equations *etc*) has enabled us to better understand what is going on. This reaction also involves acid-base phenomena and, therefore, illustrates the level to which you should be striving to apply the full breadth of your knowledge of chemistry.

The single BrO_3^- ion, single Br^- ion and two $[H^+]$ ions[l] involved in the **slowest step** (the rate-determining step) of our reaction form bromic(I) acid[m] (HOBr) and bromic(III) acid (HO_2Br):

$$BrO_3^- + Br^- + 2\,H^+ \rightarrow HOBr + HO_2Br \quad \text{(slow)}$$

In the products, bromine still has a high oxidation state (not its usual -1 or zero), so we can expect both HOBr and HO_2Br to be oxidants: they oxidise the remaining four Br^- ions in two **fast** reaction steps:

$$HO_2Br + 2\,Br^- + 2\,H^+ \rightarrow HOBr + Br_2 + H_2O \quad \text{(fast)}$$

$$2\,HOBr + 2\,Br^- + 2\,H^+ \rightarrow 2\,Br_2 + 2\,H_2O \quad \text{(fast)}$$

Notice how the second of these two reactions has been 'doubled up' (2 HOBr reacting with 2 Br^- *etc*). This is because there are two HOBr molecules to remove: one formed in the initial, slow reaction involving BrO_3^- and Br^- and another in the subsequent fast reaction between

---

[l] Strictly speaking, we should refer to these numbers as the mole ratios of the ions.

[m] You may not have heard of bromic(I) acid, but you will have encountered its cousin, chloric(I) acid, HOCl (which in the Λ-level textbooks in often given the formula HClO). This is another example of how the core material covered in the textbook can crop up in less familiar situations – you may also encounter iodic(I) acid as a reaction intermediate.

HO_2Br and Br^-. By balancing the reaction 'intermediates' HOBr and HO_2Br (which do not appear in the overall equation), they will cancel out when we add together the three steps:

$$BrO_3^- + Br^- + 2H^+ \rightarrow H\cancel{O}Br + H\cancel{O}_2Br$$

$$H\cancel{O}_2Br + 2Br^- + 2H^+ \rightarrow H\cancel{O}Br + Br_2 + H_2O$$

$$2H\cancel{O}Br + 2Br^- + 2H^+ \rightarrow 2Br_2 + 2H_2O$$

$$BrO_3^- + 5Br^- + 6H^+ \rightarrow 3Br_2 + 3H_2O$$

Referring to the initial, slow reaction between one BrO_3^-, one Br^- and two $[H^+]$ ions as the rate-determining step is potentially misleading as it may be taken to indicate it involves a simultaneous collision between the four ions. In reality, the rate-determining 'step' in this reaction is believed to proceed through a series of individual steps, known as **elementary reactions**[n]. Initially, a proton adds to the bromate(V) ion, forming bromic(V) acid:

$$BrO_3^- + H^+ \rightleftharpoons BrO_3H$$

This is simply an acid-base equilibrium, in which BrO_3H is the conjugate acid of the base BrO_3^-. BrO_3H itself then also accepts a proton in a further acid-base equilibrium, in which it is now acting as a base (with the conjugate acid $BrO_3H_2^+$):[o]

$$BrO_3H + H^+ \rightleftharpoons BrO_3H_2^+$$

So far, we have accounted for the involvement of one BrO_3^- and two H^+ ions in the rate-determining step, explaining why the reaction is first order in BrO_3^- and second order in H^+. A *single* Br^- ion (the reaction is first order in Br^-) then reacts with $BrO_3H_2^+$ to form Br_2O_2 in the reversible reaction:

$$BrO_3H_2^+ + Br^- \rightleftharpoons Br_2O_2 + H_2O$$

---

[n] This term is not used in the A-level syllabuses, but many of the reactions we encounter do occur in a series of simpler steps, so it is a handy term for us to use. In the words of Omar Khayyam: "Great fleas have little fleas upon their backs to bite 'em", And little fleas have lesser fleas, and so *ad infinitum*."

[o] The terminology of conjugate acid-base pairs is used more in some A-level courses than others (*e.g.* those of the OCR and Edexcel boards), so may not be familiar to readers following other courses.

Finally, in the elementary reaction that is the actual slowest step in the whole process (the rate-determining step), Br_2O_2 reacts with water to generate HOBr and HO_2Br:

$$Br_2O_2 \;+\; H_2O \;\rightarrow\; HOBr \;+\; HO_2Br$$

These two species then oxidise the remaining four Br^- ions in the two fast elementary steps shown above.

Please do not be at all worried if you are overwhelmed by the complexity of these reactions. The reason we are looking at this reaction system in such detail is because it is a perfect example through which we can explore the relationships between an overall chemical reaction, its rate equation and mechanism. It is these underlying principles, rather than the details of this particular reaction system, you should be trying to grasp. Indeed, the chemical literature on the kinetics and mechanism of bromide oxidation by bromate(V) under acidic conditions is far from conclusive, with research chemists disagreeing over both the rate equation and the mechanism.[p]

If we add together all the various steps that lead up to and include the reaction between Br_2O_2 and H_2O (the 'actual' rate-determining step), it is seen how the various intermediates that do not appear in the 'overall' RDS, in the form it was first written above (as a single reaction), cancel out:

BrO_3^-	$+$	H^+	$\rightleftharpoons$	BrO_3H		(fast)
BrO_3H	$+$	H^+	$\rightleftharpoons$	$BrO_3H_2^+$		(fast)
$BrO_3H_2^+$	$+$	Br^-	$\rightleftharpoons$	$Br_2O_2 \;+\; H_2O$		(fast)
Br_2O_2	$+$	H_2O	$\rightarrow$	$HOBr \;+\; HO_2Br$		(slow)

$$BrO_3^- \;+\; Br^- \;+\; 2\,H^+ \;\rightarrow\; HOBr \;+\; HO_2Br$$

Although the 'actual' rate-determining step is the slowest one of the four shown above, involving the reaction between Br_2O_2 and H_2O, these two

---

[p] Different mechanisms appear to operate at different reactant concentrations (*e.g.* at high concentration of $[Br^-]$, the reaction becomes second order in Br^-).

species do not appear as reactants in the equation for the *overall* reaction (repeated below):

$$BrO_3^- + 5\,Br^- + 6\,H^+ \rightarrow 3\,Br_2 + 3\,H_2O$$

We have seen above how, in setting out to investigate the kinetics of a reaction, we carry out experiments to determine how the rate is affected when we vary the concentration of each reactant in turn. In this case, we would vary the concentrations of BrO_3^-, Br^- and H^+ (see **Table 1.2**), but – at least in the first instance – we would have no reason to suspect that the rate is controlled by the concentrations of Br_2O_2 and H_2O. If we were, however, to have carried out experiments into the effects of $[Br_2O_2]$ and $[H_2O]$, we might have expected the rate to be directly proportional to each, reflecting the actual or 'true' rate-determining step:

$$Br_2O_2 + H_2O \rightarrow HOBr + HO_2Br$$

The reaction would be described as first order with respect to Br_2O_2 and first order w.r.t. H_2O, with the overall second-order rate equation:

$$rate = k[Br_2O_2][H_2O]$$

The reason the reaction is *observed* to be first order in BrO_3^-, second order in H^+ and first order in Br^- is because the formation of the reacting intermediates that control the overall rate of the reaction – Br_2O_2 and H_2O, in a 1:1 ratio – requires the reaction of one BrO_3^- ion, two H^+ ions and one Br^- ion.

You may find the idea that water can have a 'concentration' to be counterintuitive. You may also be wondering how the rate of the reaction can be affected – indeed, controlled – by the concentration of water when there is so much of it about; the reaction is, after all, carried out in aqueous solution. To begin with, we will deal with the concentration of water.

The molar mass of water is 18.0 g mol^{-1}, which means that if you have 18.0 g of water then you have one mole of H_2O molecules. Since the mass of 1.00 dm^3 of water is 1000 g, in 1.00 dm^3 of water there must be 55.6 moles of H_2O (1000 g ÷ 18.0 g mol^{-1} = 55.6 mol); therefore, the concentration of pure water is 55.6 mol dm^{-3}. In a reaction taking place in aqueous solution, the concentration of H_2O will, invariably, far exceed that of the reactants dissolved in the water. Consider, for example, the reactant concentrations given in **Table 1.2** for **Experiment 1**:

$$[BrO_3^-] = 1.50 \times 10^{-3} \, \text{mol dm}^{-3}$$

$$[Br^-] = 1.00 \times 10^{-1} \, \text{mol dm}^{-3}$$

$$[H^+] = 5.00 \times 10^{-2} \, \text{mol dm}^{-3}$$

The highest of these concentrations, that of Br^-, is less than 0.2 % of the concentration of water molecules. Imagine, now, that we combine BrO_3^-, Br^- and H^+ ions in an aqueous solution at the concentrations given above. The instant we do this, *before* the reaction has started, the concentration of Br_2O_2 would be zero (none has been generated), but the concentration of H_2O, the other species involved in the actual rate-determining step, would be approximately 55.6 mol dm^{-3} (ignoring the tiny effect of the dissolved reagents). At this moment, the rate is zero simply because $[Br_2O_2]$ is zero:

rate $= k[Br_2O_2][H_2O]$

$= (1.92 \, \text{dm}^9 \, \text{mol}^{-3} \, \text{s}^{-1}) \times (0.00 \, \text{mol dm}^{-3}) \times (55.6 \, \text{mol dm}^{-3})$

$= 0.00 \, \text{mol dm}^{-3} \, \text{s}^{-1}$

Now imagine the reaction is underway and that BrO_3^-, H^+ and Br^- have reacted (in the ratio 1:2:1), generating Br_2O_2 and H_2O through the elementary reaction steps leading up to the actual rate-determining step:

$$BrO_3^- + H^+ \rightleftharpoons BrO_3H$$

$$BrO_3H + H^+ \rightleftharpoons BrO_3H_2^+$$

$$BrO_3H_2^+ + Br^- \rightleftharpoons Br_2O_2 + H_2O$$

If these three steps are combined and written as a single reaction (with appropriate cancelling), it is easier to calculate the maximum concentrations of Br_2O_2 and H_2O that could be achieved:

$$BrO_3^- + 2 H^+ + Br^- \rightarrow Br_2O_2 + H_2O$$

The 'limiting' reactant under the conditions used in **Experiment 1** is BrO_3^-, the concentration of which is 1.50×10^{-3} mol dm^{-3}. Since each mole of BrO_3^- reacts to produce one mole of Br_2O_2 (and one mole of H_2O), the maximum or *theoretical* concentration of Br_2O_2 that could be achieved is 1.50×10^{-3} mol dm^{-3}. In practice, however, Br$_2$O$_2$ would never achieve this concentration because it would react immediately with H$_2$O to produce HOBr and HO$_2$Br, thus:

$$Br_2O_2 \quad + \quad H_2O \quad \rightarrow \quad HOBr \quad + \quad HO_2Br$$

The rate of this elementary reaction, which controls the rate of the overall reaction, is of course reflected in the underlying rate equation,

$$rate \quad = \quad k[Br_2O_2][H_2O]$$

and will change constantly with the changing concentrations of Br$_2$O$_2$ and H$_2$O. We have already seen that the initial concentration of water (immediately before the reaction begins) is approximately 55.6 mol dm^{-3}. Since for each mole of Br$_2$O$_2$ generated when BrO$_3^-$, H$^+$ and Br$^-$ interact in a 1:2:1 ratio, one mole of H$_2$O is produced,

$$BrO_3^- \quad + \quad 2\,H^+ \quad + \quad Br^- \quad \rightarrow \quad Br_2O_2 \quad + \quad H_2O$$

it follows that the concentration of water in the reaction mixture will increase by 1.50×10^{-3} mol dm^{-3}. The additional water generated in this reaction is, however, miniscule compared with the amount already present. A change in [H$_2$O] from 55.556 to 55.558 mol dm^{-3} is insignificant,[q] so will not affect observed rate of the reaction; therefore, in the rate equation,

$$rate \quad = \quad k\,[H_2O][Br_2O_2]$$

[H$_2$O] is, in effect, a constant:

$$rate \quad = \quad k \times a\ constant \times [Br_2O_2]$$

As the product of two constants is itself a constant, we can combine k and [H$_2$O] into a new constant and give its numerical value and units:

$$rate \ = \ (1.92\ dm^9\,mol^{-3}\,s^{-1}) \times (55.6\ mol\,dm^{-3}) \times [Br_2O_2]$$

$$rate \ = \ (1.07 \times 10^2\ dm^6\,mol^{-2}\,s^{-1}) \times [Br_2O_2]$$

We now have a very simplified rate equation for the reaction between BrO$_3^-$ with Br$^-$ ions in the presence of acid:

---

[q] In order to show that the water formed in the reaction has *any* effect on [H$_2$O], it is necessary to work to the additional decimal places.

$$\text{rate} \quad = \quad k'[Br_2O_2]$$

where k' is a 'modified' rate constant (of value 1.07×10^2 dm^6 mol^{-2} s^{-1}) that takes into account the effect of $[H_2O]$ on the rate. By treating the concentration of H_2O as a constant, and by combining it with the true rate constant, we have 'artificially' created a first-order rate equation, in which k' is said to be a **pseudo first-order** rate constant: the reaction behaves as a first-order reaction only because the concentration of water has been made so high that changes in $[H_2O]$ no longer affect the rate.

We will see in later examples how so-called 'pseudo first-order conditions' are often used by chemists to simplify the experimental investigation and analysis of reaction rates. Although you are very unlikely to see the term 'pseudo first-order' at A-level, pseudo first-order conditions are occasionally encountered in examination questions – they are just not referred to as such. It is, however, very helpful to be able to recognise when such conditions are being used, because you will then know 'where the question is going'.

We will work through the analysis of a problem involving the use of data obtained under pseudo first-order conditions in Chapter 4, but for now we must return to the oxidation of bromide ions by bromate(V) ions in acid. Although we have derived a 'rate equation' with a pseudo first-order rate constant (rate = $k'[Br_2O_2]$), it is not strictly correct for us to call this a rate equation because Br_2O_2 does not appear as a reactant in the overall equation:

$$BrO_3^- \quad + \quad 5\,Br^- \quad + \quad 6\,H^+ \quad \rightarrow \quad 3\,Br_2 \quad + \quad 3\,H_2O$$

The concentration of Br_2O_2, which determines the *overall* rate of the reaction (rate = $k'[Br_2O_2]$), is determined *indirectly* by the concentrations of the reactants BrO_3^-, Br^- and H^+, as reflected in the 'proper' – experimentally determined or 'observed' – rate equation:

$$\text{rate} \quad = \quad k[BrO_3^-][Br^-][H^+]^2$$

The reaction is *observed* to be first order in BrO_3^-, first order in Br^- and second H^+, simply because the three react in the ratio 1:1:2 in forming Br_2O_2. Water does not appear in the rate equation because it is present in excess and, in any case, cancels out when we write out the rate-determining step as a series or elementary reactions (repeated from above):

$$BrO_3^- \ + \ H^+ \ \rightleftharpoons \ BrO_3H \qquad \text{(fast)}$$

$$BrO_3H \ + \ H^+ \ \rightleftharpoons \ BrO_3H_2^+ \qquad \text{(fast)}$$

$$BrO_3H_2^+ \ + \ Br^- \ \rightleftharpoons \ Br_2O_2 \ + \ H_2O \qquad \text{(fast)}$$

$$Br_2O_2 \ + \ H_2O \ \rightarrow \ HOBr \ + \ HO_2Br \qquad \text{(slow)}$$

$$\overline{BrO_3^- \ + \ Br^- \ + \ 2\,H^+ \ \rightarrow \ HOBr \ + \ HO_2Br}$$

Indeed, when we combine the RDS with the subsequent fast reactions (also repeated from earlier), it is seen why water is, in fact, a product of the overall reaction:

$$BrO_3^- \ + \ Br^- \ + \ 2\,H^+ \ \rightarrow \ HOBr \ + \ HO_2Br$$

$$HO_2Br \ + \ 2\,Br^- \ + \ 2\,H^+ \ \rightarrow \ HOBr \ + \ Br_2 \ + \ H_2O$$

$$2\,HOBr \ + \ 2\,Br^- \ + \ 2\,H^+ \ \rightarrow \ 2\,Br_2 \ + \ 2\,H_2O$$

$$\overline{BrO_3^- \ + \ 5\,Br^- \ + \ 6\,H^+ \ \rightarrow \ 3\,Br_2 \ + \ 3\,H_2O}$$

1.5   Closing remarks

By now it should be apparent that rate equations can reflect some very complicated, underlying chemistry. Not all reactions display such complex kinetics as the example above. Consider, for example, the nucleophilic substitution reaction between the hydroxide ion and 2-bromomethylpropane we looked at earlier. In the rate-determining step, the tertiary halogenoalkane undergoes fragmentation, forming a tertiary carbocation, to which the OH^- ion then rapidly adds:

$$(CH_3)_3CBr \ \rightarrow \ (CH_3)_3C^+ \ + \ Br^- \qquad \text{(slow)}$$

$$(CH_3)_3C^+ \ + \ OH^- \ \rightarrow \ (CH_3)_3COH \qquad \text{(fast)}$$

The rate equation for this reaction is simply,

$$\text{rate} \ = \ k\,[(CH_3)_3CBr]$$

reflecting the fact that the reaction is first order, being insensitive to changes in $[OH^-]$, and that $(CH_3)_3CBr$ is the sole reactant in the rate-determining step. In the next chapter we will see that there is, however, slightly more to this reaction than meets the eye.

CHAPTER 1 SUMMARY – the key points

- The rate of a chemical reaction is defined as the change in the concentration of a reactant or product per unit time, typically having units of mol dm^{-3} s^{-1} or mol dm^{-3} min^{-1}.

- Rates are measured from plots of reactant or product concentration against time: the steeper the curve, the higher the rate. The gradient of a tangent drawn to such a curve gives the rate at precisely the time-point at which the gradient touches the curve. Tangents drawn at various time-points give a 'live report' of the rate – in much the same way as the speedometer in a motor car gives a live report of the vehicle's changing speed.

- The initial aim in any chemical kinetics investigation is to establish how the reaction rate is affected by changing individually the concentration of each reactant. By isolating the effects of each reactant, it is possible to derive a **rate equation**, which has the general form,

$$\text{rate} \quad = \quad k[A]^{a}[B]^{b}[C]^{c}$$

where k is a proportionality constant called the **rate constant**, which applies for a stated temperature. The exponents a, b and c are the order of the reaction with respect to (w.r.t.) the reactants A, B and C, respectively.

- The sum $a + b + c$ is the **order** of the reaction. The values of a, b and c must be determined experimentally: they cannot be deduced from the ratio in which A, B and C are shown to react in the overall equation for the reaction.

- Many chemical reactions occur in a series of 'steps' (elementary reactions) which, together, constitute the reaction mechanism. The overall rate of a reaction is determined by the rate of its slowest step, which is called the **rate-determining step** (RDS). Only reactants that are involved in the RDS appear in the rate equation, where their exponents (*a*, *b* and *c etc*) reflect the ratio in which they react in the RDS.

Chapter 2

Unravelling the rate constant: The Arrhenius equation

In the previous chapter, where rate equations were introduced, we saw how they consist of two components: the concentrations of particular reactants – those involved in the RDS, raised to a power that reflects the ratio in which they participate in this step – and a rate constant, which we have so far treated simply as a proportionality factor. In the present chapter, we will explore the physical basis of the rate constant and the factors that determine its magnitude.

2.1 Not all collisions are equal

Our initial discussion will be centred around two reactions that have been encountered already. The first reaction, between the hydroxide ion and bromoethane, is relatively slow,

$$CH_3CH_2Br \ + \ OH^- \ \rightarrow \ CH_3CH_2OH \ + \ Br^-$$

whereas the second reaction, the chain-terminating reaction between two methyl radicals, is extremely fast:

$$\cdot CH_3 \ + \ \cdot CH_3 \ \rightarrow \ C_2H_6$$

We can gain an appreciation of the huge difference in the rates of these two reactions by comparing the plots of OH^- and $\cdot CH_3$ concentration against time (**Figure 2.1**). In both cases, the rate of reaction decreases with time, as indicated by the decrease in the gradient – the 'steepness' – of the curve. As you will recall from GCSE chemistry, for a chemical reaction to take place the reactants must collide with each other. During the course of a reaction, the concentrations of the reactants decreases as they are converted into products. This means that the frequency of collisions between the reactants will decrease with time. Taking the decay of methyl radicals as an example, as their concentration falls, the probability of two of them 'encountering' each other also falls. At the 2.0×10^{-4} s time-point, for example, where the concentration of methyl radicals is 5.8×10^{-8} mol dm^{-3}, the rate is 2.0×10^{-4} mol dm^{-3} s^{-1} (see Chapter 1). By the time we reach the 5.0×10^{-4} s time-point,

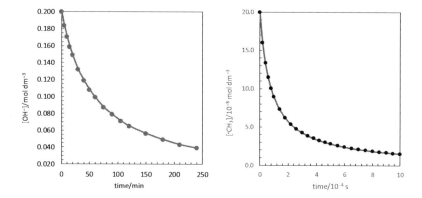

Figure 2.1 Graphs showing the decrease in reactant concentration in the reaction between bromoethane and the hydroxide ion (*left*) and during the chain-termination reaction of methyl radicals (*right*). The data plotted in the former reaction is identical to that shown in **Figure 1.1** (where the starting concentration of each reactant was 0.200 mol dm⁻³).

however, the concentration of methyl radicals has approximately halved, to 2.8×10^{-8} mol dm⁻³, and the rate is now only 4.8×10^{-5} mol dm⁻³ s⁻¹ (from the gradient of a tangent drawn at this point, which is not shown). Notice how an approximate halving of the concentration of methyl radicals has resulted in an approximate *four*-fold decrease in the rate of reaction. From this information we can deduce that the reaction is second-order with respect to the methyl radical. Recall from Chapter 1 the statement that was used to determine the order of a reaction, with respect to a given reactant, when we know how changes in the concentration of the reactant affect the rate:

> **An *x*-fold change in the concentration of the reactant causes an *x*ⁿ-fold change in the rate, where n is the order of the reaction with respect to the reactant.**

In the current example, an approximate 0.50-fold change in the concentration of the methyl radical (from 5.8×10^{-8} mol dm⁻³ to 2.8×10^{-8} mol dm⁻³) has caused the rate to change approximately 0.25-fold (2.0×10^{-4} mol dm⁻³ s⁻¹ multiplied by 0.25 is 5.0×10^{-5} mol dm⁻³ s⁻¹). Since 0.25 equals 0.50 to the power 2 (0.50^2), it follows that the order of the reaction with respect to the methyl radical is 2, leading to the rate equation:

$$rate = k \, [\text{\textbullet}CH_3]^2$$

It is very common in examination questions to be asked to determine reaction orders and rate equations when given such data (in the form of a graph or a table), therefore now is a good time to learn a little trick you can apply to simplify the process when the data you are given concerns the *decrease* in the rate of a reaction in response to a decrease in reactant concentration. The trick is to reverse the order in which you consider the data: instead of saying a 0.50-fold change in reactant concentration (*i.e.* halving it) causes a 0.25-fold change in the rate (it falls to ¼ of its original value), it is much easier to think in terms of a doubling in the concentration of the reactant causing a 4-fold increase in the rate. It is easier to 'see' this if we put the data in a mini-table:

$[\text{\textbullet}CH_3]/\text{mol dm}^{-3}$	$\text{rate}/\text{mol dm}^{-3}\,\text{s}^{-1}$
2.8×10^{-8}	4.8×10^{-5}
5.8×10^{-8}	2.0×10^{-4}

Although the concentration of methyl radicals was $5.8 \times 10^{-8}\,\text{mol dm}^{-3}$ *before* it fell to $2.8 \times 10^{-8}\,\text{mol dm}^{-3}$, the order in which the rates at these concentrations were measured does not change the fact that when $[\text{\textbullet}CH_3]$ is doubled the rate increases approximately 4-fold. In the form of our statement, we can now say: because a 2-fold change in $[\text{\textbullet}CH_3]$ causes a 2^2-fold (*i.e.* 4-fold) change in the rate, the reaction is 2nd order with respect to $[\text{\textbullet}CH_3]$.

We now have rate equations for both the nucleophilic substitution and radical chain-termination reactions:

$$rate = k \, [\text{OH}^-][CH_3CH_2Br]$$

$$rate = k \, [\text{\textbullet}CH_3]^2$$

Note that both reactions are second order *overall*: the former is first order with respect to both OH^- and CH_3CH_2Br, whereas the latter is second order with respect to $\text{\textbullet}CH_3$. This tells us that each reaction involves a direct bimolecular collision: between OH^- and CH_3CH_2Br in one case and between two $\cdot CH_3$ radicals in the other. Each reaction, then, occurs in a single step; there are no separate, slow and fast steps. Despite this similarity, there is a huge difference in the rates of the two reactions – such that the former had to be monitored over a period measured in tens of minutes, if not hours, whereas the latter was largely complete within a thousandth of a second.

To try to explain the difference in the rates of the two reactions, let's first consider the difference in reactant concentrations. In the nucleophilic substitution, the starting concentrations of the hydroxide ion and bromoethane were both 0.200 mol dm^{-3}, which is one-million times higher than the starting concentration of methyl radicals (20.0×10^{-8} mol dm^{-3}). We might reasonably expect, therefore, that at these reactant concentrations, OH^- ions and CH_3CH_2Br molecules will be colliding with each other much more frequently than pairs of $\cdot CH_3$ radicals. The fact that the rate of the nucleophilic substitution reaction is, nevertheless, much *slower* than that of the radical chain-termination reaction suggests that the frequency of collisions between the reactants is not the most important factor in determining of the rate of a chemical reaction: a far more important factor is the energy with which the collisions occur. Only collisions that involve a minimum amount of energy, the so-called **activation energy**, can lead to a chemical change – namely, a reaction. You will no doubt already have some idea what is meant by the activation energy of a reaction from your GCSE studies; this concept will, however, be developed in the present chapter.

2.2 'Unpacking' the rate constant

The huge difference in the rates of our two reactions is reflected in their rate constants. At a value of 6.20×10^{10} dm^3 mol^{-1} s^{-1},[r] the rate constant for the combination of methyl radicals is over 10^{13} times greater than that for the reaction of the hydroxide ion with bromoethane, 1.43×10^{-3} dm^3 mol^{-1} s^{-1} (see Chapter 1). Although you will certainly need to understand the underlying reasons why rate constants can have such vastly different values, it can also be very helpful, when comparing

---

[r] A range of values for this rate constant is seen in the chemical literature (depending on the solvent and other factors). The value given here is representative for the combination of methyl radicals in the gas phase.

reaction rates, to think of the rate constant of a reaction as *the rate of the reaction when the concentrations of all the reactants equals 1.00 mol dm^{-3}*. Thus, for the reaction of the hydroxide ion with bromoethane (with units omitted for clarity):

$$\text{rate} = k \times 1 \times 1$$

$$= k$$

since $k = 1.43 \times 10^{-3} \ \text{dm}^3 \ \text{mol}^{-1} \ \text{s}^{-1}$, we have (with the appropriate cancelling of units):

$$\text{rate} = 1.43 \times 10^{-3} \ \text{mol dm}^{-3} \ \text{s}^{-1}$$

From this 'default' value, it is now a simple matter to predict the effects of changes in reactant concentration on the rate of the reaction. It can be seen, for example, how doubling the concentration of bromoethane causes the rate to double (the reaction is first order with respect to CH_3CH_2Br):

$$\text{rate} = k \times 1 \times 2$$

$$= 2k$$

$$= 2.86 \times 10^{-3} \ \text{mol dm}^{-3} \ \text{s}^{-1}$$

In contrast, doubling the concentration of methyl radicals results in a 4-fold increase in their rate of combination. Thus, when [·CH$_3$] is 1.00 mol dm^{-3}:

$$\text{rate} = k \times 1^2$$

$$= k$$

$$= 6.20 \times 10^{10} \ \text{mol dm}^{-3} \ \text{s}^{-1}$$

but when [·CH$_3$] equals 2.00 mol dm^{-3}:

$$\text{rate} = k \times 2^2$$

$$= 4k$$

$$= 2.48 \times 10^{11} \ \text{mol dm}^{-3} \ \text{s}^{-1}$$

To achieve a 4-fold increase in the rate of the nucleophilic substitution reaction, we could, for example, double both [OH$^-$] and [CH$_3$CH$_2$Br] to 2.00 mol dm^{-3}:

$$\text{rate} = k \times 2 \times 2$$

$$= 4k$$

$$= 5.72 \times 10^{-3} \text{ mol dm}^{-3} \text{ s}^{-1}$$

This approach can be very helpful when tackling examination questions where, although it is unlikely you will be dealing with whole-number concentrations, the principle is the same: whatever the concentration of – to use the present example – the methyl radical in one experiment, doubling its concentration in another experiment will result in a 4-fold increase in the rate.

We will now turn our attention to the physical basis of the rate constant and explore the factors that determine its magnitude. The symbol used for the rate constant, k, looks innocent enough, but scrape beneath the surface and you will discover a concoction of complicated factors that must be addressed in any attempt to account for the value of the rate constant for a given reaction at a stated temperature. The three 'ingredients' that go into determining the rate constant are:

(i)   a steric factor, P, which takes into account the orientation of the collisions between the reacting species in the rate-determining step;

(ii)   a collision frequency factor, Z, which, as you might expect, concerns the frequency with which the reacting species collide in the RDS; and

(iii)   the Boltzmann term, $e^{-(E_a/RT)}$, which concerns the energy with which the reacting species collide in the RDS,

where E_a is the activation energy of the reaction, R is the gas constant (8.314 J K^{-1} mol^{-1}) and T is temperature on the kelvin scale. If you are studying mathematics at A-level, you may already be familiar with exponential terms and the constant, e (Euler's number), which has a value of 2.71828 to 5 decimal places. If e is new to you, there is no need to worry – all will be explained in due course.

The rate constant is directly proportional to each of these factors, being the product:

$$k = P \times Z \times e^{-(E_a/RT)}$$

P and Z, the steric and collision frequency factors, are usually rolled into one, to make a single constant, A, known variously as the pre-exponential factor, the Arrhenius constant or the collision frequency factor (which can be confusing because this term also refers to Z):

$$k = A \times e^{-(E_a/RT)}$$

This is the **Arrhenius equation**, which evolved from a similar equation proposed by the Swedish chemist Svante Arrhenius in 1889. Arrhenius arrived at his equation following the analysis of eight sets of published data on the effect of temperature on reaction rates. The authors of the individual studies he analysed had each proposed their own equations to describe the temperature dependence of the rate constant. Indeed, the equation published by the Dutch chemist Jacobus Henricus van't Hoff was essentially the same as that later proposed by Arrhenius. The Swede, however, was the first to argue that the equation was applicable to all chemical reactions.

Although the Arrhenius equation was proposed on the basis of empirical evidence (it 'fitted' the experimental data), it has since been demonstrated that the equation can be derived from a purely theoretical consideration of the factors that affect reaction rates – the three 'ingredients' listed above. At A-level, it is necessary to have *some* understanding of the theoretical basis of the Arrhenius equation – as well as the ability to obtain the activation energy and Arrhenius constant of a given reaction from experimental data.[s] We will begin with the former, namely the development of your knowledge of 'collision theory' – first introduced at GCSE level – to form the basis of your understanding of the theory behind the Arrhenius equation.

2.3 Collision theory – reaching the transition state

According to the **collision theory** of reaction rates, as taught in schools at GCSE, AS and A-level, a reaction can take place only when the

---

[s] Not all of the examination boards require candidates to be able to determine the Arrhenius constant from data (*e.g.* the current Edexcel specification).

reactant particles – whether molecules, atoms, ions or radicals – collide with each other with a minimum amount of energy, which we saw above is referred to as the activation energy (E_a). The magnitude of the activation energy reflects the amount of energy needed to break the chemical bonds necessary for product formation to be initiated.

In its most elementary form, collision theory assumes that reactant particles behave as fast-moving, tiny spheres, travelling in straight lines (randomly, in all directions), and do not interact with each other until the moment of their impact. The nearest we can get to this hypothetical situation is the gas state, in which the particles are so small compared with the average distances between them, we can ignore any forces between the particles. Although most gases do come close to this 'ideal' behaviour (assumed when using the ideal gas equation), this is certainly not the case for reactions in the liquid state.

Consider, for example, the reaction between CH_3CH_2Br and the OH^- ion: even in a very dilute solution, where the distances between the reactant particles are large, there must come a point, immediately before the moment of collision, at which the partial positive charge on carbon-1 of a bromoethane molecule begins to attract the negatively-charged hydroxide ion, thereby 'bringing on' a collision. Without this attractive force suddenly kicking-in, the two species might not collide at all, but just pass each other by in a near-miss. Thus, if we were to attempt to estimate the frequency with which CH_3CH_2Br molecules and OH^- ions collide in solution, based only on their concentration and average velocity[t], we would be underestimating the true value.

Another assumption made in collision theory is that the reactant particles behave as hard, non-compressible spheres – rather like tiny ball-bearings – that are converted into products 'instantaneously' upon collision. In practice, when reactants collide, their kinetic energy is channelled into a form of potential energy in which chemical bonds are stretched and deformed before finally breaking. The difference between the type of interaction assumed to occur in collision theory and what actually takes place is not unlike the difference between a pair of snooker balls colliding and a tennis ball hitting the ground: whereas snooker balls are hard and are therefore not visibly compressed or deformed in a

---

[t] Velocity is speed in a stated direction, so is it the term used when the direction in which an object is travelling is of importance.

collision, the softer tennis ball becomes slightly compressed ('squashed') – momentarily changing shape – before rebounding.

Now imagine a pair of gaseous reactant molecules *before* they interact – still too far away from each other for any intermolecular forces to be acting between them. The molecules will be moving very quickly, so they possess kinetic energy. As you will recall from GCSE physics, the kinetic energy (*KE*) of an object is a function of both its mass, *m* (in kg), and its velocity, *v* (in m s^{-1}):

$$KE = \tfrac{1}{2}\, m\, v^2$$

The kinetic energy of a single oxygen molecule,[u] for example, travelling at 482 m s^{-1} (the average speed of O_2 molecules at 25 °C) is 6.17×10^{-21} joules (J):

$$KE = \tfrac{1}{2} \times (5.314 \times 10^{-26}\ \text{kg}) \times (482\ \text{m s}^{-1})^2$$

$$= \tfrac{1}{2} \times (5.314 \times 10^{-26}\ \text{kg}) \times 232324\ \text{m}^2\ \text{s}^{-2}$$

$$= 6.17 \times 10^{-21}\ \text{kg m}^2\ \text{s}^{-2}$$

The units that result from this calculation, kg m^2 s^{-2}, are an example of *base units*. Energies are usually given in units of the joule, which is a *derived unit* (1 J = 1 kg m^2 s^{-2}). Unless you are also studying physics at A-level, you need not be overly concerned with the distinction and conversion between base units and derived units; however, being able to obtain the final units in a calculation from those of the input values is an important skill that should be practised by all chemists; notice, for example, how the units of the velocity (m s^{-1}) have been squared along with its value:

$$(482\ \text{m s}^{-1})^2 = 482\ \text{m s}^{-1} \times 482\ \text{m s}^{-1} = 232324\ \text{m}^2\ \text{s}^{-2}$$

Although we have now strayed somewhat into the realm of physics, students following the current AQA and OCR Salters courses do need to be familiar with (and able to use) the above equation for kinetic energy when answering questions on time-of-flight mass spectrometry.

---

[u] The mass of a single O_2 molecule in grams is calculated by dividing the molar mass of O_2 (the mass of one mole of molecules, 32.0 g mol^{-1}) by Avogadro's number. This must then be converted into kilograms for the calculation of its *KE*.

The kinetic energy of our O_2 molecule (6.17×10^{-21} J) may seem very small, but if we had a mole of such molecules, with an average speed of 482 m s^{-1}, their combined KE would be 3.72 kJ.[v] When a pair of molecules collide head-on, they suddenly lose kinetic energy as their velocity decreases. Energy, of course, is never lost, rather it is converted into another form of energy, which in this case is potential energy. Imagine a trampolinist falling at high speed towards a trampoline: the moment contact is made with the trampoline, the trampolinist rapidly loses velocity and therefore kinetic energy. As the trampoline stretches, the trampolinist's kinetic energy is being transformed into elastic potential energy. At the lowest point on the trampoline, when it has stretched to its maximum and the trampolinist is momentarily stationary, the trampolinist will have zero kinetic energy but maximum elastic potential energy. The conversion of this potential energy back into kinetic energy is responsible for flinging the trampolinist back into the air.

Something very similar happens when a pair of molecules collide, but they fly apart again only if the energy of the collision is less than the activation energy. If the molecules have enough kinetic energy, the bending and deformation of their chemical bonds – during which the molecules' kinetic energy is being converted into potential energy – will result in their breaking and (may) lead to a reaction. It is the equivalent of the trampoline being stretched so far that it breaks. The difference, however, is that whereas when the trampoline breaks it is irreversible (and unfortunate for the trampolinist), a chemical bond that breaks may immediately reform again; therefore, a chemical reaction is not inevitable.

Oxygen molecules are of course colliding constantly in the air, but because the chemical bond in an O_2 molecule is strong (it is a double covalent bond), very few collisions will be of sufficient energy to result in bond breaking. Similarly, the even stronger, triple bonds between the atoms in N_2 molecules mean that very few of their collisions result in bond breaking. At the very high temperatures achieved in vehicle engines, however, these molecules do have sufficient kinetic energy to result in a reaction when they collide. The reaction between oxygen and nitrogen molecules in vehicle engines results in the formation of nitrogen monoxide, which is a free radical, as shown in the equation:

---

[v] $(6.17 \times 10^{-21}$ J$) \times (6.022 \times 10^{23}$ mol$^{-1}) = 3.72 \times 10^3$ J mol^{-1}.

$$O_2 \;+\; N_2 \;\rightarrow\; 2\;{}^{\bullet}NO$$

The subsequent reaction of ${}^{\bullet}NO$ with oxygen results in the formation of nitrogen dioxide (${}^{\bullet}NO_2$), which is a more reactive radical and is responsible for much of the respiratory and other health problems associated with vehicular pollution (along with particulate matter):

$$2\;{}^{\bullet}NO \;+\; O_2 \;\rightarrow\; 2\;{}^{\bullet}NO_2$$

In considering the events that occur *during* the collision between a pair of molecules, we have strayed into the realms of a more sophisticated model of reaction kinetics, known as **transition state theory**. Although transition state theory is not usually addressed at A-level, one or two of the concepts it introduces are of relevance to your studies. In the AQA course, for example, mention is made of the transition state of a chemical reaction. (See also the various definitions of activation energy given below.)

The transition state is the chemical species[w] that is formed when a reaction reaches its point of highest energy; the energy of the transition state determines the activation energy for the reaction.[x] This is best illustrated by returning to the reaction between the hydroxide ion and bromoethane:

$$1\;CH_3CH_2Br \;+\; 1\;OH^- \;\rightarrow\; CH_3CH_2OH \;+\; Br^-$$

where,

$$\text{rate} \;=\; k\,[OH^-]^1\,[CH_3CH_2Br]^1$$

It was mentioned above how the rate equation for this reaction, in which the power to which the concentration of each reactant is raised (in this case 1) is the same as its 'balancing number' in the overall equation (added to emphasise this point), tells us that the reaction occurs in a single step, involving a direct collision between OH^- and CH_3CH_2Br.

---

[w] For reasons that will be explained in Chapter 3, do not refer to the transition state as a 'reaction intermediate', which has an entirely different meaning.

[x] As we will also see in Chapter 3, reactions that occur in more than one step (*i.e.* they involve a rate-determining step and one or more much faster steps) have a transition state for each, individual step.

This is often shown by the curly-arrow mechanism we saw in Chapter 1:

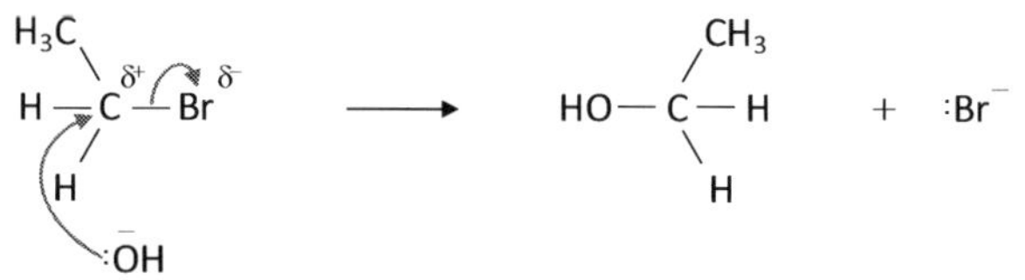

Notice how formation of the bond between the hydroxide ion and the carbon atom and the breaking of the carbon-to-bromine bond occur simultaneously. In the transition state, it is as though we have 'frozen' the reaction at its halfway point, in which the OH group and Br atom are simultaneously attached to the carbon atom by half-formed and half-broken bonds, respectively. In the following diagram, 'wedges' have been used to show how the geometry of the bonds around carbon-1 'flips' as the reaction proceeds through the transition state:

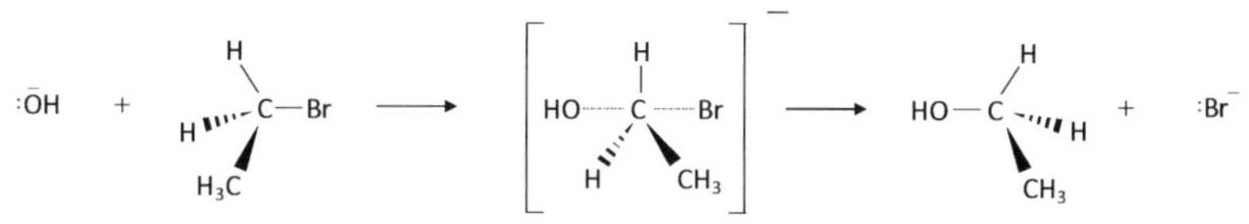

transition state

For the transition state to be formed, the combined kinetic energy of the OH$^-$ ion and CH_3CH_2Br molecule – during their rapid approach towards each other – must be high enough to bring their potential energy up to that of the transition state in the collision. Before the moment of impact, however, the reactants are considered to have zero potential energy, possessing only kinetic energy. *During* a 'successful' collision (*i.e.* one that results in the reaction), as the reactants lose velocity, their kinetic energy is being channelled into potential energy, which reaches a maximum value in the transition state. This can be shown using a **reaction profile diagram** (or energy profile diagram): a graph of the energy changes during a reaction, where the activation energy corresponds to the amount of potential energy the reactants must gain to reach the transition state and thereby react.

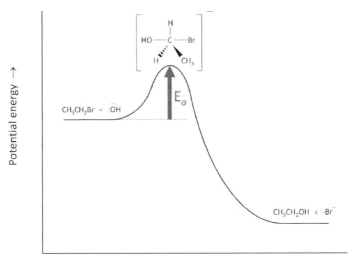

Potential energy →

Progress of reaction →

Figure 2.2   Reaction profile for the reaction between the hydroxide ion and bromoethane. The transition state is the species present when the reaction reaches its position of highest energy. The arrow, labelled E_a, is the activation energy, which for this reaction is 89.5 kJ mol^{-1}.

Shown in **Figure 2.2** is the energy profile for the reaction between the hydroxide ion and bromoethane, which has an activation energy of 89.5 kJ mol^{-1}. As the potential energy of the reactants is assumed to be zero before the collision, you can think of 89.5 kJ mol^{-1} as either: (i) the absolute value of the *potential* energy of the transition state; or (ii) the *change* in potential energy that occurs during a successful collision.

In A-level textbooks, the *y*-axis in reaction-profile diagrams is typically labelled 'energy' or 'enthalpy', rather than 'potential energy' (as in **Figure 2.2**). It is, however, only by being aware of the subtle distinction discussed here that we can understand why the activation energy for a given reaction is constant: it does not vary with changes in temperature. When a reaction mixture is heated, the reactant particles move around faster, reflecting the increase in their kinetic energy. Although the *total* energy of the reactants – the sum of their kinetic and potential energies – increases, there is no change in their potential

54

energy, which is taken to be zero because the reactant particles are considered to be so far apart from each other that no forces are acting between them (until the moment of the collision, that is). If the y-axis of the reaction profile represented 'energy', then, increasing the temperature would mean we would have to place the initial energy level of the reactants higher up on the axis to reflect the increase in their total energy (kinetic plus potential). The amount of energy the reactants would then need to gain to reach the transition state would be reduced (the red arrow would be shorter in **Figure 2.2**), giving the impression that the activation energy of a reaction decreases with increases in temperature.

Once the transition state has been achieved, it may not necessarily lead to product formation. In the nucleophilic substitution reaction shown in **Figure 2.2**, the transition state could simply break apart to form the reactants again, rather than the products. The situation is analogous to that of someone attempting to push a boulder over a hump before it can be released to roll down a hillside: from its position at the top of the hump (where it has maximum gravitational potential energy), the boulder may, instead, simply roll back to its starting position.

The activation energy makes its appearance in the Arrhenius equation within the Boltzmann term, $e^{-(E_a/RT)}$, which computes, at a given temperature, the fraction of collisions between the reactants that occur with an energy equal to or exceeding the activation energy. In a reaction mixture containing CH_3CH_2Br molecules and OH^- ions at at temperature of, for example, 75 °C (348 K), the fraction of the collisions that occur with an energy of at least 89.5 kJ mol^{-1} is 3.62×10^{-14}:

$$\text{fraction} = e^{-\left(\frac{89500 \text{ J mol}^{-1}}{8.31 \text{ J K}^{-1} \text{ mol}^{-1} \times 348 \text{ K}}\right)}$$

$$= e^{-30.95}$$

$$= 3.62 \times 10^{-14}$$

The reason e (Euler's number), which we encountered earlier, appears in the Boltzmann term will be explained in detail shortly; for the time being, all you need to remember is that it is simply a constant and that you should use the $e^{\blacksquare}$ key on your calculator when functions of the type $e^{-30.95}$ appear in calculations.

Notice how the activation energy has been entered in J mol^{-1}, rather than kJ mol^{-1}. These units then cancel with the J mol^{-1} within the units of the gas constant, R, leaving K^{-1}, which is cancelled by the temperature unit, K. Notice also that, although the final answer has been given appropriately to 3 significant figures (*i.e.* not more than smallest number of significant figures of the values entered into the calculation[y]), the e$^{-30.95}$ term has not been rounded to 3 figures. You should always carry 'too many' significant figures *during* a calculation, rounding only in the final step. Had e$^{-30.95}$ been rounded to 3 significant figures 'too early' (to e$^{-31.0}$), the final answer would have been different (3.44×10^{-14}). *To avoid rounding errors, round to the appropriate number of significant figures only at the last possible stage in your calculations.*

The fraction calculated above is simply a proportion, meaning '3.62 $\times$ 10^{-14} in 1' collisions occur with an energy of at least 89.5 kJ mol^{-1}. You may find it easier to think of this as a percentage: 3.62×10^{-14} in 1 is the same ratio as 3.62×10^{-12} in 100. Whichever way you look at it, the percentage of collisions occurring with sufficient energy to generate the transition state (3.62×10^{-12} %) is extremely small, leading one to wonder how on earth do the hydroxide ion and bromoethane ever manage to react with each other at a perceivable rate! The solution to this dilemma lies in the fact that, although only a tiny percentage of collisions occur with the minimum energy needed for a reaction to take place, there are so many collisions occurring every second (greater than 10^{10}), the reaction is able to proceed at a meaningful rate of knots. Before turning our attention to the Arrhenius constant (A), which factors the collision frequency into the calculation of the rate constant, we will take a closer look at the Boltzmann term and how it relates to the Boltzmann distribution.

2.4 The Boltzmann and Maxwell-Boltzmann distributions

The Boltzmann distribution is a graph showing the distribution of the kinetic energies of the individual particles (usually molecules) in a gas at a stated temperature. **Figure 2.3** shows the Boltzmann distribution for

---

[y] Using the convention that the zeros after a non-zero number count as significant figures, the activation energy written as 89500 J mol^{-1} is considered to be to 5 significant figures. Any ambiguity here can be avoided using standard form: 8.95×10^4 J mol^{-1} is the value to 3 significant figures.

a gas at 800 K. The red arrow shows how we would 'read off' the fraction of molecules that have a kinetic energy of, for example, 10.0 kJ mol^{-1}, which is 0.0116. Thus, in a population of 10,000 molecules at 800 K, we would expect 116 to have a kinetic energy of 10.0 kJ mol^{-1} (*i.e.* 1.16 % of them).

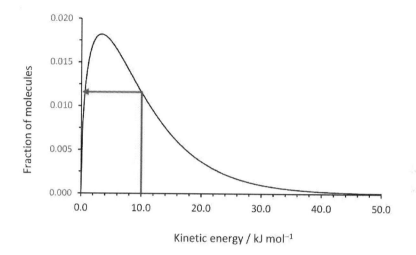

In order to appreciate the connection between the Boltzmann distribution and the $e^{-(E_a/RT)}$ term in the Arrhenius equation, it is helpful to have a slightly deeper understanding of how the Boltzmann distribution is calculated. Rather than giving the fraction of molecules that have a specific amount of kinetic energy, such as 10.0 kJ mol^{-1}, the graph shown in **Figure 2.3** tells us, in fact, the proportion of molecules with a kinetic energy within the range 10.00 to 10.25 kJ mol^{-1}. We should, therefore, modify the above statement and say that 0.0116

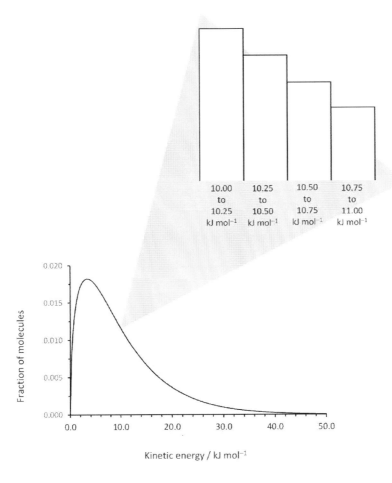

Figure 2.4   An enlarged view of a tiny section of the Boltzmann distribution shown in **Figure 2.3**. Upon magnification, the curve is revealed to be a histogram, in which each column gives the fraction of molecules with a kinetic energy within each interval of 0.25 kJ mol^{-1}. See text for details.

(1.16 %) is the fraction of molecules with a kinetic energy between 10.00 and 10.25 kJ mol^{-1}.

In constructing a Boltzmann distribution, we are not restricted to a range of 0.25 kJ mol^{-1}. Instead, any energy interval – of, say, 0.50 kJ mol^{-1} (10.00 to 10.50 kJ mol^{-1}), or 1.00 kJ mol^{-1} (10.00 to 11.00 kJ mol^{-1}) – may be used. Of course, if we continue to increase the interval size, the graph will change from being a curve to a histogram. The reason the distribution shown in **Figure 2.3** has the appearance of a smooth curve is because the energy interval used in the underlying calculations is very small. It is helpful, however, to be aware that if we were to magnify ('zoom in' on) a section of the curve, it would be seen to consist of a series of columns, in which the height of each represents the fraction of molecules with a kinetic energy within the chosen interval (**Figure 2.4**).

You may have noticed that the identity of the gas in **Figures 2.3 and 2.4** has not been stated. This is because *all gases display the same distribution of their kinetic energy at a given temperature:* it makes no difference whether the gas is hydrogen, oxygen, ethane, or any other gas. Recall that the molecules in an 'ideal' gas are considered to have zero potential energy: they possess only kinetic energy. This is because the distances between the molecules are so large (compared with their sizes), there are assumed to be no intermolecular forces acting between them.

Since the kinetic energy of an object is determined by both its velocity and its mass, when we say that a molecule of hydrogen, for example, has a particular kinetic energy, its mass has already been taken into account: it has been factored into the calculation of its kinetic energy. We can calculate the kinetic energy of a *single* H_2 molecule with a kinetic energy of, for example, 10.0 kJ mol^{-1}, by dividing the energy of one mole of H_2 molecules by Avogadro's constant:

$$KE = \frac{10000 \text{ J mol}^{-1}}{6.022 \times 10^{23} \text{ mol}^{-1}}$$

$$= 1.66 \times 10^{-20} \text{ J}$$

It may be helpful to think of this calculation in reverse: the total kinetic energy possessed by one mole of H_2 molecules, each with kinetic energy of 1.66×10^{-20} J, is simply 6.022×10^{23} 'lots' of 1.66×10^{-20} joules:

$$KE = (1.66 \times 10^{-20} \text{ J}) \times (6.022 \times 10^{23} \text{ mol}^{-1})$$

$$= 10.0 \times 10^{3} \text{ J mol}^{-1}$$

$$= 10.0 \text{ kJ mol}^{-1}$$

Similarly, the mass (m) of a single H_2 molecule is obtained by dividing its molar mass by the Avogadro constant,

$$mass \quad = \quad \frac{2 \text{ g mol}^{-1}}{6.022 \times 10^{23} \text{ mol}^{-1}}$$

$$= \quad 3.32 \times 10^{-24} \text{ g}$$

from which we find the velocity (v) of a H_2 molecule that has a kinetic energy of 1.66×10^{-20} joules to be just over 3 km per second:

$$KE \quad = \quad \tfrac{1}{2} \, m \, v^2$$

$$1.66 \times 10^{-20} \text{ J} \quad = \quad \tfrac{1}{2} \times (3.32 \times 10^{-27} \text{ kg}) \times v^2$$

$$v \quad = \quad 3.16 \times 10^3 \text{ m s}^{-1}$$

Compare this with the velocity of an oxygen molecule with the *same* kinetic energy:

$$KE \quad = \quad \tfrac{1}{2} \, m \, v^2$$

$$1.66 \times 10^{-20} \text{ J} \quad = \quad \tfrac{1}{2} \times (5.31 \times 10^{-26} \text{ kg}) \times v^2$$

$$v \quad = \quad 7.91 \times 10^2 \text{ m s}^{-1}$$

It is seen that, due to its greater mass, the O_2 molecule needs to travel at only 0.791 km per second to possess the same kinetic energy as a molecule of hydrogen travelling at 3.16 km per second.

It follows, then, that whereas the **Boltzmann distributions of energy** are identical for all molecules at a given temperature, the so-called **Maxwell distributions of speeds** for different gases are different – even at the same temperature. **Figure 2.5** shows the Maxwell distributions of speed for oxygen and hydrogen gases at 273 K (0 °C), 800 K (527 °C) and 1600 K (1327 °C).

The Maxwell distribution of speeds was first reported by James Clerk Maxwell in 1860. His reasoning involved the application of probability theory to the behaviour of gases. Ludwig Boltzmann subsequently arrived at the same description, basing his derivation on the *energy* distribution of the particles in gases. For this reason, the Maxwell

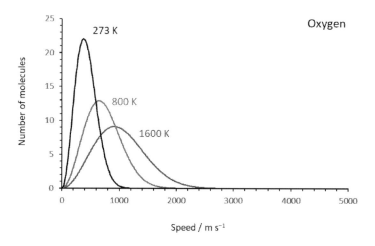

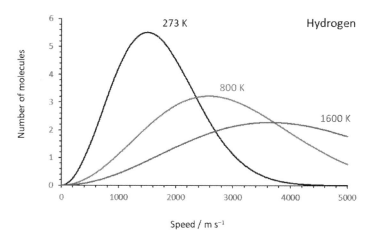

Figure 2.5 Maxwell distributions of speed for oxygen (*upper*) and hydrogen (*lower*) molecules at temperatures of 273 K, 800 K and 1600 K. Each plot has been normalised to a population of 1000 molecules. See text for details.

distribution of speeds is often referred to as the **Maxwell-Boltzmann distribution of molecular speeds** (or velocities), thereby reflecting the contribution of both scientists. In the various A-level chemistry courses, these terms can be used rather loosely: whereas the OCR specification (including the Chemistry B Salters course) correctly uses the term Boltzmann distribution when referring to molecular energies, the Edexcel and AQA boards use the term 'Maxwell-Boltzmann

61

distribution of molecular energies'. Strictly speaking, Maxwell should be mentioned only when referring to molecular speeds, either alone or with Boltzmann.

Just as the Boltzmann distribution is essentially a histogram with very narrow column widths, so too is the Maxwell distribution, the difference being that in the latter the height of each column represents the fraction of molecules whose speed falls within the specified interval. The distributions shown in **Figure 2.5** were calculated using an interval of 10 m s^{-1}. At 273 K, for example, the fraction of H_2 molecules with a speed in the range 1200 to 1210 m s^{-1} is 0.00504. The distributions shown in **Figure 2.5** have each been applied ('normalised') to a population of 1000 molecules. Thus, the plot for H_2 at 273 K shows that, in a population of 1000 molecules, we would expect to find approximately 5 travelling at a speed between 1200 and 1210 m s^{-1} (0.00504 × 1000 = 5.05). Using the formula $KE = \frac{1}{2} m v^2$, and taking the mass of a H_2 molecule to be 3.32 × 10^{-27} kg (see above), we can say that the kinetic energy of each of our five H_2 molecules is within the range 2.39 × 10^{-21} to 2.43 × 10^{-21} J.

You may, perhaps, be wondering, if all one thousand molecules are at the same temperature, why do they not all have the same speed and therefore the same kinetic energy? The answer lies in the fact that the molecules in a gas are constantly undergoing random collisions: each molecule is considered to be moving in a straight line until it collides with another; if a single molecule is struck simultaneously by two or three others on the same side, it will be propelled to a greater speed, so its kinetic energy will increase.[z] If a gas is left at a given temperature for enough time, these random collisions will have the effect of 'distributing' the total kinetic energy of the gas unevenly between the individual molecules, resulting in a Boltzmann distribution. (The gas is now said to have reached thermodynamic equilibrium.)

The A-level courses tend to focus on energy distributions – rather than their underlying speed distributions. You are expected to know, for example, how to 'read' a Boltzmann distribution. We have seen already how to read-off from a Boltzmann distribution the fraction (or number)

---

[z] As you will have learnt in GCSE physics, the total momentum of objects involved in a collision is conserved, so if one object's momentum (its mass multiplied by its velocity) decreases, that of one or more of the others must increase.

of molecules in a population that have a particular kinetic energy; indeed, we saw how this reading is really telling us the fraction of molecules with kinetic energies within a specified, narrow range. The **most probable energy** in a population of molecules at a given temperature is simply the energy that is possessed by the greatest fraction of molecules: it is read off from the 'peak' of the curve, as shown in **Figure 2.6**.

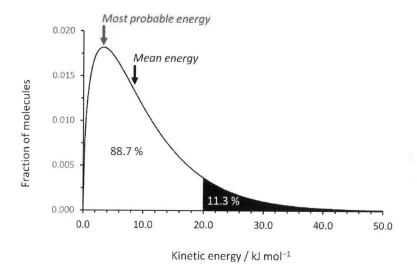

Figure 2.6   Boltzmann distribution of kinetic energy across a population of gaseous molecules at a temperature of 800 K. The upper arrow shows how we would read off the most probable energy in the population of molecules, which at this temperature is 3.25 kJ mol^{-1}. The filled area – representing 11.3 % of the 'area under the curve' – shows the proportion of molecules with energy of at least 20.0 kJ mol^{-1}. See text for details.

Note, however, that the most probable energy of the molecules (3.25 kJ mol^{-1} at 800 K) is *not* the same as the average (mean) energy across the population, which is approximately 8.5 kJ mol^{-1} at 800 K.[aa]

At this point, it is helpful to recall that the Boltzmann distribution is really a histogram, which – because it consists of a very large number of columns – has the appearance of a continuous curve. If you were to add

---

[aa] At A-level, you are not required to be able to calculate these values.

up the heights of all the columns in **Figure 2.6** (each equal to the fraction of molecules with energy within the 10.0 kJ mol^{-1} range it spans), the total would be 1. In other words, the fraction of the molecules having a kinetic energy between zero and infinity is 1.00 (100 % of the population).[bb] For this reason, if you are asked to sketch a Boltzmann distribution in the examinations, your curve must begin at the origin – the number of molecules having zero kinetic energy is zero – and continue to infinity, thereby including all possible kinetics energies (however unlikely): in other words, it is important that your curve never quite touches the *x*-axis.[bb]

When adding up the heights of all the columns that went into producing the curve plotted in **Figure 2.6**, what we are really doing is calculating the total 'area under the curve'. (Students taking A-level mathematics will already have recognised this process as integration.) The area under the curve measured over a defined range of kinetic energies gives us the fraction – proportion or percentage – of molecules with energies within the defined range. Thus, the filled area in **Figure 2.6**, representing 11.3 % of the total 'area under the curve',[cc] corresponds to the fraction of the molecules whose kinetic energy is equal to or greater than 20.0 kJ mol^{-1}. In other words, at 800 K, 11.3 % of the molecules in a sample of any gas have a kinetic energy of at least 20.0 kJ mol^{-1}.

The standard A-level chemistry textbooks tell you (in one way or another) that the area under the curve, extending from the activation energy to infinity, corresponds to the fraction of molecules that have the minimum amount of energy needed for a reaction take place. Thus, for a reaction with an activation energy of 20.0 kJ mol^{-1}, at 800 K, you might expect that – because 11.3 % of the molecules possess at least 20.0 kJ mol^{-1} of kinetic energy at this temperature (**Figure 2.6**) – 11.3 % of the molecules will collide with sufficient energy for a reaction to take place. This, however, is not quite correct. It is necessary here to appreciate the distinction between the kinetic energies of the *individual* molecules in a

---

[bb] The Boltzmann distribution shown in **Figure 2.6** was computed for kinetic energies up to 50.0 kJ mol^{-1}, but strictly speaking – in order to include all possible energies – the *y*-axis should be extended to infinity. For this reason, the area under the curve (obtained by adding up all the column heights) is slightly less than 0.997, rather than 1.00. Let's not lose any sleep over the 0.3 % of molecules whose kinetic energy is greater than 50.0 kJ mol^{-1}!

[cc] As a fraction (*i.e.* out of 1.00), this would be 0.113.

population (shown by the Boltzmann distribution) and the energy of the impacts when *pairs* of molecules collide. The energy of these impacts depends not only on the speed of each molecule (and therefore its kinetic energy), but also the angle at which the molecules collide. If two molecules collide head-on, all their energy will be channelled into the collision, thereby achieving the maximum potential energy upon impact. If, however, the molecules experience only a glancing, side-on collision, the impact energy will be lower, possibly failing to reach the minimum value needed for a reaction to take place; in this scenario, the molecules will simply be deflected and continue on their merry way. A Boltzmann distribution of molecular energies takes no account of the direction in which the individual molecules are travelling, but this is a crucial consideration in assessing the energies of the collisions between the molecules.

Fortunately, the Boltzmann term – which is built into the Arrhenius equation – takes care of this for us: the fraction of *collisions* that occur with an energy equal to or greater than the activation energy is given by $e^{-(E_a/RT)}$. Thus, for a reaction with an activation energy of 20.0 kJ mol^{-1}, the probability of a collision occurring with at least the minimum amount of energy for a reaction to be possible is 0.0494:

$$\text{probability} \;=\; e^{-\left(\dfrac{20000 \text{ J mol}^{-1}}{8.31 \text{ J K}^{-1} \text{ mol}^{-1} \times 800 \text{ K}}\right)}$$

$$=\; e^{-3.01}$$

$$=\; 0.0494$$

This value is referred to as a probability (0.0494 in 1) because the Boltzmann distribution is a statistical description of the behaviour of a very large population of molecules (a description of 'crowd behaviour'): as the population is so large, it is possible to equate probability with proportion or fraction. Thus, we would expect 4.94 % of the collisions taking place at any moment to occur with an energy of at least 20.0 kJ mol^{-1}, which is somewhat less than the 11.3 % 'area under the curve' given in **Figure 2.6.** Rather than saying, then, that the area under the curve 'to the right of' a boundary drawn at the activation energy equals the fraction or number of molecules that have the minimum amount of energy needed to react, we should be saying that this area is *proportional* to the number of *collisions* that occur at an energy that equals or exceeds the activation energy: the greater the area under the curve to the right of the activation energy, the greater will be the number of collisions occurring at or above this energy. The distinction here is not subtle (4.94 as opposed to 11.3 %), yet it appears to be far from clear in the standard A-level textbooks.

2.5 Temperature – the key factor in determining reaction rate

As shown in **Figure 2.7**, changing the temperature has a huge effect on the proportion of collisions that occur with energies at or above the activation energy. For example, doubling the temperature from 800 K to 1600 K causes a 4.5-fold increase in the proportion of collisions with an energy of at least 20.0 kJ mol^{-1} (0.0494 to 0.222). This is the reason why rate constants (and hence reaction rates) respond 'out of all proportion' to changes in temperature.

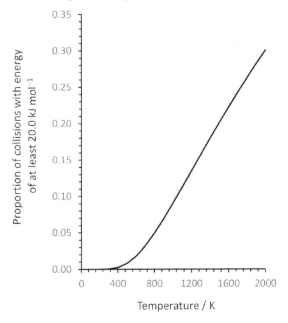

Figure 2.7 A graph showing the effect of temperature on the proportion of collisions in a population of molecules occurring with an energy of at least 20.0 kJ mol^{-1}. The *y*-axis can also be taken to be the probability of a collision occurring with at least an energy of 20.0 kJ mol^{-1}. See text for details.

The Boltzmann component of the Arrhenius equation, $e^{-(E_a/RT)}$, is of the mathematical form e^{-x}, with E_a/RT written in place of x. Euler's number, e, appears in formulae where one variable – in this case, the fraction of collisions occurring with an energy of at least E_a – changes by an increasingly large amount in response to changes in another variable (in this case, temperature). The table below shows how e^{-x} diminishes rapidly as x is increases (try calculating these values for yourself). Notice, in particular, how the *extent* by which e^{-x} diminishes depends on the size of x. Thus, doubling x from 1 to 2 causes e^{-x} to decrease by around two-thirds (from 0.368 to 0.135), but doubling x from 5 to 10 causes e^{-x} to decrease by more than 99 %. Variables that respond in this manner – and can, therefore, be described mathematically using a term of the form e^{-x} – are said to display exponential decay.

x	e^{-x}
1	0.368
2	0.135
5	0.00674
10	0.0000454
25	0.0000000000139

Anything that causes 'x' to be large – such as a high activation energy – will result in a disproportionately large *decrease* in the value of e^{-x}, the fraction of collisions occurring with an energy of at least the activation energy.

Consider two reactions, one with an activation energy of 20.0 kJ mol^{-1} and another of 100.0 kJ mol^{-1}. The values of $e^{-(E_a/RT)}$, calculated over a wide range of temperatures, are given in **Table 2.1**. To develop confidence in your mathematical skills, try these calculations for yourself.[dd] There is no substitute for practise! Note the enormous impact increasing the activation energy from 20.0 to 100.0 kJ mol^{-1} has

---

[dd] With an activation energy of 20.0 kJ mol^{-1} (20000 J mol^{-1}), you should obtain the following values of $-E_a/RT$ (to 4 significant figures and using R = 8.314 J K^{-1} mol^{-1}): −8.210 (at 293 K), −7.686 (at 313 K), −7.224 (at 333 K) *etc.*

on the fraction of collisions occurring with an energy of at least E_a. At 293 K, for example, this drops by a factor of almost two-hundred-trillion (1.8×10^{14})! Notice, in particular, that ***the higher the activation energy of a reaction, the more sensitive the term $e^{-(E_a/RT)}$ is to changes in temperature***. Thus, whereas a modest increase in temperature from 293 to 313 K results in only a 1.7-fold increase the fraction of collisions occurring with an energy of at least 20.0 kJ mol^{-1}, the same temperature change causes a 13.8-fold increase in the fraction of collisions occurring with an energy of at least 100.0 kJ mol^{-1}. This is the reason why rate constants and – consequently – reaction rates are so extremely sensitive to changes in temperature. If you understand this, you understand the Arrhenius equation.

Temperature / K	$e^{-(E_a/RT)}$ (Fraction of collisions occurring with an energy of at least E_a)	
	$E_a = 20.0$ kJ mol^{-1}	$E_a = 100.0$ kJ mol^{-1}
293	2.72×10^{-4}	1.49×10^{-18}
313	4.60×10^{-4}	2.05×10^{-17}
333	7.29×10^{-4}	2.06×10^{-16}
353	1.10×10^{-3}	1.59×10^{-15}
553	1.29×10^{-2}	3.58×10^{-10}
1053	1.01×10^{-1}	1.09×10^{-5}

Table 2.1 Data illustrating the effect of temperature on the proportion of molecular collisions occurring at or above the activations energy for two hypothetical reactions, one with an activation energy of 20.0 kJ mol^{-1}, the other 100.0 kJ mol^{-1}. Notice how the effect of increasing the temperature is more pronounced in the reaction with the higher activation energy.

The $e^{-(E_a/RT)}$ term within the Arrhenius equation, then, determines the fraction of collisions that occur with the minimum amount of energy required for a reaction to take place; in other words, the activation energy. The other key factor that determines the rate of a reaction is, of course, the collision frequency. Thus, we calculated above that, at 800 K, the fraction of the collisions between molecules that occur with an energy of at least 20.0 kJ mol^{-1} is '0.0494 in 1', which is 4.94 % – but 4.94 % of what? We need to know how many collisions are occurring each second, so we can calculate 4.94 % of this to obtain the number of collisions each second that result in a reaction.

2.6 Collisions – their frequency and geometry

When a rate equation is used to work out the rate of a reaction for a given set of reactant concentrations, the collision frequency is being factored into the calculation in *two* places: firstly, in the reactant concentrations (the higher the reactant concentrations, the higher the collision frequency); and, secondly, in the Arrhenius constant (A), otherwise known as the collision frequency factor or – because it appears in the equation in front of the exponential term – the pre-exponential factor. This is best explained using a real example. Consider the addition of hydrogen to ethene to give ethane:

You may recall from your GCSE and Year-12 studies that alkenes are hydrogenated in the presence of a nickel[cc] catalyst (*e.g.* during the 'hardening' of vegetable oils). Although this reaction is highly exothermic ($\Delta H^\circ \approx -140$ kJ mol^{-1}), meaning the products are much lower in energy than the reactants, in the absence of a catalyst the reaction is extremely slow. There are two reasons for this: (i) the activation energy is very high (180 kJ mol^{-1}); and (ii) the pre-exponential factor, A, is relatively small (1.24×10^6 dm^3 mol^{-1} s^{-1} at 628 K) – for most reactions, A is around 10^{11}.[ff] By inputting these values into the Arrhenius equation, we obtain the rate constant at 628 K:

[cc] Other catalysts may also be used, including palladium and platinum.
[ff] The pre-exponential factor has the same units as the rate constant. The hydrogenation of ethene is a second-order reaction in the absence of a catalyst (rate = $k[H_2][C_2H_4]$), so the units of A are dm^3 mol^{-1} s^{-1}.

$$k = (1.24 \times 10^6 \text{ dm}^3 \text{ mol}^{-1} \text{ s}^{-1}) \times e^{-\left(\frac{180{,}000 \text{ J mol}^{-1}}{8.31 \text{ J K}^{-1} \text{ mol}^{-1} \times 628 \text{ K}}\right)}$$

$$= (1.24 \times 10^6) \times e^{-34.5}$$

$$= (1.24 \times 10^6 \text{ dm}^3 \text{ mol}^{-1} \text{ s}^{-1}) \times 1.05 \times 10^{-15}$$

$$= 1.30 \times 10^{-9} \text{ dm}^3 \text{ mol}^{-1} \text{ s}^{-1}$$

Due to its very small rate constant, the rate of this reaction is so slow that, to all intents and purposes, it is considered not to 'go' at all.

Catalysts speed up chemical reactions by providing an alternative 'pathway' from reactants to products; in other words, the reaction proceeds by a different mechanism, involving a transition state that is lower in energy, resulting in a lower activation energy. In the case of hydrogen addition to an alkene, the reaction takes place on the surface of the metal catalyst, to which the gaseous reactant molecules become temporarily attached ('adsorbed').

The catalysts of reactions that take place in aqueous solution are themselves typically dissolved in the solution, along with the reactants. For example, many such reactions are catalysed by aqueous hydrogen ions or metal ions. We will examine how these catalysts work in greater detail in later chapters, where we will look in detail at some of the reactions that come up frequently in examination questions (*e.g.* the acid-catalysed halogenation of propanone).

Although the direct reaction between hydrogen and ethene in the absence of a catalyst is of little practical relevance, it can serve as a relatively simple (albeit theoretical) model system through which we can explore the physical basis of the pre-exponential factor, A. As stated above, two factors account for the exceptionally low rate constant for this reaction: a high activation energy and a small pre-exponential factor – what you might call a 'double whammy'. The reason for the high activation energy is that the reactants must form a highly-unstable transition state, consisting of a four-membered ring structure, which is some 180 kJ mol^{-1} in energy above that of the reactants (see **Figure 2.8**).

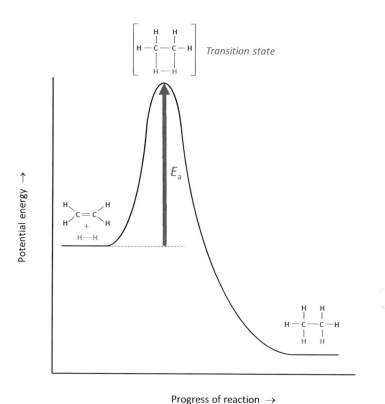

Figure 2.8 Reaction profile for the hydrogenation of ethene. The formation of a highly unstable species in the transition state, consisting of a four-membered ring structure, accounts for the high activation energy of the reaction (180 kJ mol⁻¹).

The rate equation for the reaction is:

$$\text{rate} = k \times [H_2][CH_2CH_2]$$

Since the rate constant, k, is given by (the Arrhenius equation),

$$k = A\, e^{-(E_a/RT)}$$

we can make the substitution:

$$\text{rate} = A\, e^{-(E_a/RT)} \times [H_2][CH_2CH_2]$$

The Arrhenius constant (A) allows for the fact that, although the collision frequency is *proportional* to the reactant concentrations, it is *not numerically equal to* the product of their concentrations. In a gaseous

71

reaction mixture containing H_2 and CH_2CH_2 at concentrations of, for example, 0.10 and 0.20 mol dm^{-3}, respectively, multiplying $[H_2]$ by $[CH_2CH_2]$ gives 0.020 mol^2 dm^{-6}. To convert this into a collision frequency, it must be multiplied by another factor, A, the value of which is unique for each reaction. Changing the order of the terms in the previous expression helps illustrate this point:

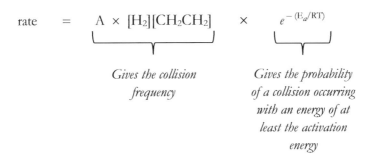

$$\text{rate} \quad = \quad \underbrace{A \times [H_2][CH_2CH_2]}_{} \quad \times \quad \underbrace{e^{-(E_a/RT)}}_{}$$

Gives the collision
frequency

Gives the probability
of a collision occurring
with an energy of at
least the activation
energy

To understand the role of the Arrhenius constant in deriving the collision frequency for a given reaction, imagine we have two sealed containers of volume 1 dm^3, one containing helium and the other argon.[gg] If both containers are at a temperature of 293 K and a pressure of 100 kPa, they must each contain 2.47×10^{22} atoms (0.0411 moles), calculated using the ideal gas equation.[hh] Although both containers hold the same number of gaseous atoms, confined to the same volume, the number of collisions occurring in each container will not be the same. When asked in which container there will be more collisions per second, most students opt intuitively for the one containing argon – argon atoms are larger than helium atoms, so surely they bump into each other more often if they are confined to the same volume? They are, however, wrong!

Since both populations of atoms are at the same temperature, they will have the same total kinetic energy (distributed across their 2.47×10^{22} atoms, according to the Boltzmann distribution). As the kinetic

---

[gg] Although these gases are unreactive, they were selected for use in this example because they occur as single atoms, which allows us to 'approximate' their behavior to that of a population of fast-moving, colliding spheres.
[hh] For guidance on the ideal gas equation, and other routine calculations that underpin A-level chemistry, see: *Understanding Calculations in AS and First Year A-level Chemistry: A guide for the utterly confused* (Mark Burkitt, 2017).

energy of an object is a function of both its velocity and its mass ($KE =$ ½ mv^2), it follows that – to have the same total kinetic energy as the argon atoms – the lighter, helium atoms must be travelling at a higher average speed. (See also **Figure 2.5**, above, which shows the Maxwell distributions of speed of hydrogen and oxygen molecules at various temperatures: at any given temperature, the H_2 molecules have a greater average velocity than the heavier O_2 molecules).

At a temperature of 293 K, the average speed of helium atoms (1352 m s^{-1}) is, in fact, some three times greater than that of their heavier, argon counterparts (428 m s^{-1}).[ii] Thus, although argon atoms are considerably larger than helium atoms, it is the fact that the latter are travelling so much faster that accounts for their greater collision frequency: in our 1 dm^3 sealed container containing 2.47×10^{22} helium atoms at a temperature of 293 K and a pressure 100 kPa, there will be a total of 9.93×10^9 collisions every second; in the equivalent container holding the same number of argon atoms, there will be 5.38×10^9 collisions each second. Despite their larger size, then, the lower average speed of the argon atoms accounts for the fact, that under the conditions stated above, their collision frequency is lower than that of the smaller (but faster) helium atoms.

The above conclusion reflects, of course, the 'ideal gas' model used in the physical sciences to explain the properties of gases, in which the sizes of the individual gas particles (whether atoms or molecules) are considered to be insignificant compared with the average distances between them. From the average speed and collision frequency of helium atoms at 293 K and 100 kPa, it can be shown that, on average, a helium atom will travel a distance of approximately 140 nm (1.4×10^{-7} m) before colliding with another atom: that is over 500-times its diameter. Similarly, under the same conditions, an atom of argon (travelling more slowly) will cover some 80 nm before colliding with another atom, which is almost 250-times its diameter.

The sole purpose of the various figures – speeds, collision frequencies and distances – quoted in the previous two paragraphs is to illustrate, using real examples, the behaviour of the individual particles

---

[ii] The 'speeds' quoted here are root-mean-square (rms) speeds, which are used when accounting for the properties of gases. Students who are studying A-level physics may already be familiar with the concept of the rms speed, but it is not required for A-level chemistry.

(whether atoms or molecules) in a gaseous population: it is to convey the idea that the particles are travelling at very high speeds; undergo an enormous number of collisions every second; and yet cover very large distances (relative to their diameter) between collisions. At A-level, of course, you are not expected to be able to calculate the various values I have quoted[ii]. Indeed, the Arrhenius constant (A) takes care of all this for you: when multiplied by the reactant concentrations, it gives the collision frequency within a population of gas particles at a stated temperature and concentration.

This can be illustrated by returning to the reaction between hydrogen and ethene. At 628 K, the value of A for this reaction is 1.24×10^6 dm^3 mol^{-1} s^{-1}. At the H_2 and CH_2CH_2 concentrations chosen above (0.10 and 0.20 mol dm^{-3}, respectively), a collision frequency of 2.48×10^4 mol dm^{-3} s^{-1} is obtained:

$$\text{collision frequency} = (1.24 \times 10^6 \ dm^3 \ mol^{-1} \ s^{-1}) \times [H_2][CH_2CH_2]$$

$$= (1.24 \times 10^6 \ dm^3 \ mol^{-1} \ s^{-1}) \times 0.10 \ mol \ dm^{-3}$$

$$\times 0.20 \ mol \ dm^{-3}$$

$$= 2.48 \times 10^4 \ mol \ dm^{-3} \ s^{-1}$$

If we now multiply this collision frequency by the fraction of collisions that occur with an energy at or above the activation energy, which is 1.05×10^{-15} at 628 K (see above), we arrive at the *rate* of the reaction under these conditions:

$$\text{rate} = (2.48 \times 10^4 \ mol \ dm^{-3} \ s^{-1}) \times (1.05 \times 10^{-15})$$

$$= 2.60 \times 10^{-11} \ mol \ dm^{-3} \ s^{-1}$$

As mentioned earlier, without a suitable catalyst, this reaction is too slow to be of any practical relevance; it does, however, serve as a relatively straightforward example of a reaction that illustrates clearly the phenomena being discussed here.

[ii] As mentioned previously, students following courses that cover time-of-flight mass spectrometry (*e.g.* AQA) are expected to be able to carry out calculations involving the formula for kinetic energy ($KE = \frac{1}{2} mv^2$).

In calculating the reaction rate above, the collision frequency and probability of a collision occurring with an energy of at least the activation energy were calculated separately and then multiplied:

$$\text{rate} \quad = \quad A[H_2][CH_2CH_2] \quad \times \quad e^{-(E_a/RT)}$$

The $e^{-(E_a/RT)}$ term has no units; it is simply a probability (a '1.05×10^{-15} in one chance'). The units of rate arise from the collision frequency component in the calculation: multiplying the units of A by those of $[H_2]$ and $[CH_2CH_2]$ gives, after appropriate cancelling, $mol\ dm^{-3}\ s^{-1}$:

$$\cancel{dm^3}\ \cancel{mol}^{-1}\ s^{-1}\ (A) \ \times\ \cancel{mol}\ dm^{-3}\ ([H_2]) \times\ mol\ \cancel{dm}^{-3}\ ([CH_2CH_2])$$

$$= \ mol\ dm^{-3}\ s^{-1}\ (\text{rate})$$

Rather than inputting a separate collision frequency factor and Boltzmann term each time we need to calculate the rate of a reaction, we use instead the rate constant, k, which is simply the product of the two – 'rolled onto one', as it were. The same numbers and units are being multiplied, only in a different order, allowing us to use the same rate constant for any given set of reactant concentrations:

$$\text{rate} \quad = \quad A\ e^{-(E_a/RT)} \ \times\ [H_2][CH_2CH_2]$$

$$= \quad k\ \times\ [H_2][CH_2CH_2]$$

As mentioned above, the Boltzmann term has no units, so the units of the rate constant are given by the units of A, the collision frequency factor. For a second-order reaction, which involves a collision between two species in the rate-determining step, A (and therefore, k), has units of $dm^3\ mol^{-1}\ s^{-1}$. After multiplying these units by those of the concentrations of the two reactants in the second-order rate equation, we are left – after cancelling down – with $mol\ dm^{-3}\ s^{-1}$, which are the units of rate.

It follows, then, that the units of for a first-order reaction are s^{-1} and those for a third-order reaction are $dm^6\ mol^{-2}\ s^{-1}$. For example, for the reaction between nitrogen monoxide and oxygen,

$$2\ NO_{(g)}\ +\ O_{2(g)}\ \rightarrow\ 2\ NO_{2(g)}$$

the rate equation is,

$$\text{rate} = \ k[NO]^2[O_2]$$

75

where the units of A (and therefore k) are $dm^6 \, mol^{-2} \, s^{-1}$. Multiplying these units by those of the reactant concentrations given in the rate equation gives, after cancelling, the correct units of rate:

$$mol \, dm^{-3} \, s^{-1} \, (rate) \ =$$

$$dm^6 \, mol^{-2} \, s^{-1} \, (k) \ \times \ mol^2 \, dm^{-6} \, ([NO_2]^2) \times \ mol \, dm^{-3} \, [O_2]$$

It was mentioned earlier in this chapter that the Arrhenius equation was originally proposed following the analysis of data from a series of experiments in which the effect of temperature on the rate constant was investigated. It was also mentioned that, in more recent years, the equation has been derived on purely theoretical grounds, which involved calculating the collision frequency between the reactant molecules at a given temperature and concentration (or pressure). Collision frequency factors obtained from experimental data – using a procedure that will be illustrated below – are typically smaller than the theoretical values calculated using the speeds and diameters of the molecules concerned. For the reaction between hydrogen and ethene, for example, the actual value of A $(1.24 \times 10^6 \, dm^3 \, mol^{-1} \, s^{-1})$ – determined from experimental data and used in the calculation above – is considerably smaller than the theoretical value of $7.37 \times 10^{11} \, dm^3 \, mol^{-1} \, s^{-1}$. In many cases, however, the theoretical value is not so far from the experimental value. For example, for the combination of two methyl radicals,

$$\cdot CH_3 \ + \ \cdot CH_3 \ \rightarrow \ C_2H_6$$

the theoretical and experimental values are, respectively, 1.1×10^{11} and $2.4 \times 10^{10} \, dm^3 \, mol^{-1} \, s^{-1}$. Similarly, for the reaction,

$$2 \, NOCl \ \rightarrow \ 2 \, NO \ + \ Cl_2$$

the theoretical value of A is $5.9 \times 10^{10} \, dm^3 \, mol^{-1} \, s^{-1}$ and the experimental value is $9.4 \times 10^9 \, dm^3 \, mol^{-1} \, s^{-1}$. The explanation as to why the actual values of the collision frequency factor – those obtained from experimental data – are generally lower than their theoretical counterparts lies in the fact that, for a reaction to occur, the reactant molecules must not only collide with a minimum amount of energy, they must collide with a specific geometry. In the case of the reaction between hydrogen and ethene, for example, only a tiny proportion of the collisions $(1.68 \times 10^{-6}$ in 1) collisions occur in an orientation that enables the highly unstable transition-state species to be formed (**Figure 2.9**).

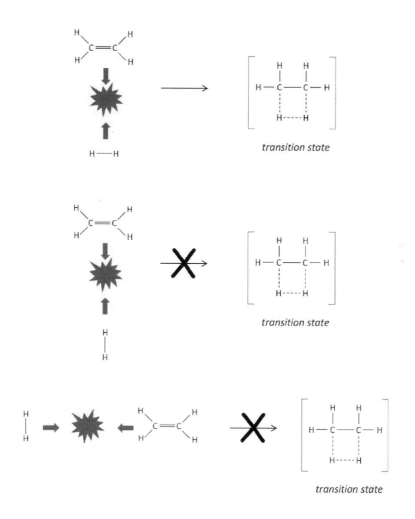

Figure 2.9   In the reaction between hydrogen and ethene, only those collisions that occur at or above the activation energy – *and* with the correct geometry – lead to the formation of the transition state and, thereby, result in a reaction.

The reason for the much better agreement between the theoretical and experimental values of A in the vast majority of reactions (*e.g.* between two methyl radicals or two NOCl molecules) lies in the fact that the geometrical constraints on the collisions in such reactions are far less demanding, allowing a greater proportion of collisions to result in the formation of the transition state.

Earlier in this chapter, the three 'ingredients' that go into determining the size of the rate constant were introduced, namely, the steric factor, P, the collision frequency factor, Z, and the Boltzmann term, $e^{-(E_a/RT)}$. The steric factor, P, is simply a 'correction factor' by which the collision frequency factor, Z, is multiplied to take into account the fact that not all of the collisions occur in an orientation that allows a reaction to take place, irrespective of whether or not they occur with an energy that equals or exceeds the activation energy. The 'uncorrected' collision frequency factor, Z, then, corresponds to the theoretical value of A, calculated solely on the basis of the speeds and diameters of the reactant molecules. In the reaction between hydrogen and ethene, P has a value of 1.68×10^{-6}, which means that only '1.68×10^{-6} in 1 collisions' (0.000168 %) occur in an orientation that allows the transition state to be formed. Multiplying the theoretical, calculated collision frequency factor by this value gives A, the Arrhenius constant:

$$A = Z \times P$$

$$= (7.37 \times 10^{11} \, dm^3 \, mol^{-1} \, s^{-1}) \times 1.68 \times 10^{-6}$$

$$= 1.24 \times 10^6 \, dm^3 \, mol^{-1} \, s^{-1}$$

Notice that P has no units; it is simply a probability. As mentioned earlier, the steric factor for this reaction is particularly small because hydrogen and ethene must collide in a very specific geometrical arrangement for the highly-unstable, four-membered ring structure of the transition state to be formed (see **Figures 2.8** and **2.9**). In contrast, the steric factor for the chain-termination reaction between a pair of methyl radicals is 0.22, meaning that 22 % of collisions occur in an orientation that allows the reaction to proceed:

$$A = (1.1 \times 10^{11} \, dm^3 \, mol^{-1} \, s^{-1}) \times 0.22$$

$$= 2.4 \times 10^{10} \, dm^3 \, mol^{-1} \, s^{-1}$$

In the usual form of the Arrhenius equation, the collision frequency factor appears as A, *i.e.*, after it has been 'corrected' by the steric factor.

At GCSE level, and perhaps also in Year-12, you will have been told that increasing the temperature increases the rate of a reaction because it results in an increase in the proportion of collisions occurring with an energy equal to or greater than the activation energy *and* the collision frequency. Although you will have been told that it is the former of these two that has the greater impact on the rate, you will nevertheless have learned that, when the temperature is increased, particles move faster and therefore collide more frequently. You may by now be wondering, therefore, why the Arrhenius equation makes no allowance for the effect of temperature on the Arrhenius constant: the temperature value we enter into the Arrhenius equation is used only to calculate the fraction of the collisions that occur with an energy at or above the activation energy; it is not used to 'adjust for temperature' the collision frequency factor, A, which occurs as a separate term in the equation, independent of the Boltzmann term:

$$k \; = \; A \; \times \; e^{-(E_a/RT)}$$

The truth of the matter is that the collision frequency factor *is* affected by temperature but, *by such a small amount compared with the Boltzmann term,* it is treated as though it were temperature independent. Thus, when calculating the *theoretical* value of the collision frequency factor for a given reaction, it is necessary to take into account temperature (as well as the size and mass of the reactant particles). When this is carried out for the reaction between hydrogen and ethene at 628 K, a value of 7.37 $\times$ 10^{11} dm^3 mol^{-1} s^{-1} is obtained. This is the value of Z that was given above and multiplied by the steric factor (1.68×10^{-6}) to give A (1.24×10^6 dm^3 mol^{-1} s^{-1}). The corresponding value of A calculated after doubling the temperature to 1256 K – and multiplying by the same steric factor, which is not affected by changes in temperature – is 1.75×10^6 dm^3 mol^{-1} s^{-1}.

We see, then, that doubling the temperature has resulted in a 41 % increase in the Arrhenius constant, from 1.24×10^6 to 1.75×10^6 dm^3 mol^{-1} s^{-1}. Let's now look at the effect of the same temperature change on the contribution made by the Boltzmann term, $e^{-(E_a/RT)}$, to the rate constant. Increasing the temperature from 628 to 1256 K causes this to increase from 1.05×10^{-15} (see above) to 3.26×10^{-8}, which represents an increase of over 3 billion per cent. It is clear, then, that increasing the temperature increases the rate constant, *k*, primarily through its effect on the Boltzmann term, *i.e.*, by causing a huge increase in the proportion of collisions that occur with energy at or above the activation energy. The values of A, $e^{-(E_a/RT)}$ and *k* at 628 and 1256 K are summarised in

79

Table 2.2, along with the calculated rates when hydrogen and ethene are present at concentrations of 0.10 and 0.20 mol dm^{-3}, respectively.

Temperature /K	A /dm^3 mol^{-1} s^{-1}	$e^{-(E_a/RT)}$	k /dm^3 mol^{-1} s^{-1}	Rate* /mol dm^{-3} s^{-1}
628	1.24×10^6	1.05×10^{-15}	1.3×10^{-9}	2.6×10^{-11}
1256	1.75×10^6	3.26×10^{-8}	5.7×10^{-2}	1.1×10^{-3}

Table 2.2　Summary of data illustrating how increasing the temperature increases the rate constant of a reaction (between hydrogen and ethene) primarily through its effect on the Boltzmann term, $e^{-(E_a/RT)}$. Although the collision frequency factor, A, also increases with temperature, its effect on the rate constant is insignificant compared with impact of the Boltzmann term. *When $[H_2]$ = 0.10 mol dm^{-3} and $[CH_2CH_2]$ 0.20 mol dm^{-3}.

We can now summarise by saying that a change in temperature affects the value of a rate constant primarily through its effect on the Boltzmann component of the Arrhenius equation. The collision frequency factor, A, is treated as being temperature independent: when using the Arrhenius equation to calculate a rate constant at given temperature, the temperature value is entered within the $e^{-(E_a/RT)}$ term; but the value of A entered is the same for all temperatures.

It is also worth noting here that increasing the temperature has a much greater effect on the rate of a reaction than increasing the concentration of a reactant by the same factor – a point occasionally seen in examination questions.[kk] We see from in **Table 2.2** , for example, that with hydrogen and ethene at concentrations of 0.10 and 0.20 mol dm^{-3}, respectively, doubling the temperature from 628 to 1256 K causes the rate to increase some 42 million-fold (from 2.6×10^{-11} to 1.1×10^{-3} mol dm^{-3} s^{-1}). If, instead, we were to double the concentration of hydrogen (or ethene), but keep the temperature at 628 K, the rate would

---

[kk] See, for example, Question 1.4 in the AQA 2015 Specimen Paper 2.

increase only by a factor of two. Thus, whereas doubling the concentration of hydrogen doubles only the frequency of collisions that occur with an energy of at least the activation energy,[ll] doubling the temperature results in a much greater increase in the frequency of such collisions, as reflected in the huge increase in the $e^{-(E_a/RT)}$ term.

2.7 Obtaining the activation energy and Arrhenius constant from experimental data

We have seen how – using the theoretical concepts of collision frequency, collision geometry and activation energy – the Arrhenius equation provides us with the means to rationalise the value of a given rate constant. We now turn our attention to the practical application of the equation – to the determination of an activation energy (E_a) and Arrhenius constant (A) from experimental data. At A-level, you are expected to be fully familiar these procedures,[mm] described and explained here in the following worked example, which concerns the decomposition of ethanal to methane and carbon monoxide:

The reaction has been found to be second order with respect to ethanal, obeying the following rate equation:

$$\text{rate} = k\,[CH_3CHO]^2$$

[ll] This reaction is first-order with respect to both [H_2] and [CH_2CH_2]. If a reaction is second-order in a particular reactant, doubling the concentration of the reactant will, of course, result in a 4-fold increase in the rate, but this effect is still much smaller than that of an equivalent increase in the temperature.
[mm] Although students following the AQA, OCR A and Chemistry B (Salters) specifications are expected to be able to calculate both activation energies and Arrhenius constants (A) from experimental data, the specification for the Edexcel course states only that students need to be able to calculate activation energies. It is, however, very important to check your specification regularly, as changes do occur.

This suggests the reaction involves the collision of two CH_3CHO molecules. **Figure 2.10** shows a plot of the rate constant against temperature. To obtain the activation energy and collision frequency factor from this plot, we could attempt to 'fit' the line-of-best-fit to the Arrhenius equation. This might involve entering pairs of temperature and rate constant values into the equation and 'trying out' different values of the two constants, E_a and A, until we 'hit upon' values that reproduce the line-of-best-fit seen in **Figure 2.10**. Using computer simulation software, this is a relatively straightforward procedure, but it is more common to use a 'manual' method, in which the Arrhenius equation is first converted into a logarithmic form.

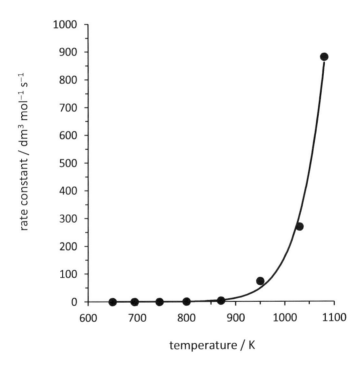

Figure 2.10   A plot showing the effect of temperature on the value of the rate constant for the second-order decomposition of ethanal to methane and carbon monoxide.

In examination questions, you are usually given two versions of the Arrhenius equation:[nn] the version we have used above and the logarithmic version, which on the Data Sheet for the OCR board is given as:

$$\ln k = -E_a/RT + \ln A$$

If you are taking mathematics A-level, you will find it a simple matter to obtain this equation by taking natural logarithms of the Arrhenius equation. If you are unfamiliar with logarithms, there is no need to be alarmed, as the all the mathematical procedures you need to be able to carry out will be explained in due course.

In its logarithmic version, the Arrhenius equation is of the form $y = mx + c$. In other words, it is the equation of a straight line. This can be seen more clearly when the question is 'opened up' and written as:

$$\ln k = -\frac{E_a}{R} \times \left[\frac{1}{T}\right] + \ln A$$

$$y = m \times x + c$$

This opened-up form of the equation is similar to the version that has been given in past examination questions set by the Edexcel board.[oo] Whatever logarithmic version of the Arrhenius equation you are given, you must be able to write it in the form shown above – either from memory or by recasting the single-line version of the equation ($\ln k = -E_a/RT + \ln A$).

As the equation is of the form $y = mx + c$, plotting $\ln k$ on the y-axis against $1/T$ on the x-axis will give a straight line. The gradient (m) of the line-of-best-fit is equal to $-E_a/R$ and $\ln A$ is equal to the value of $\ln k$ when $1/T$ is zero (*i.e.* where the line intercepts the y-axis). In **Table 2.3**, it is shown how the temperature and rate constant values are processed for plotting in this form. (The graph shown earlier, in **Figure**

[nn] At the time of writing, the AQA and Edexcel boards give the equations 'when required', whereas the OCR board (including the Salters B Specification) give the equations on the Data Sheet. You must, however, check the current policy with your own board.

[oo] The current Edexcel specification does not require you to obtain A, so the ln A term in the equation has been given as a constant in past questions (see, for example, Question 9 from the 2018 Paper 2).

2.10, was plotted using the 'unprocessed' data from *columns 1* and *4* in the table.) Although '1 divided by the temperature' (*column 2*) must be plotted on the *x*-axis, it is common to multiply these values by 1000, thereby avoiding the need to show lots of decimal places (or use standard form) on the axis. In practice, then, we plot '1000 divided by the temperature' (*column 3*) on the *x*-axis, not forgetting to take this into account later when measuring the gradient. If you are already familiar with logarithms, *column 5* will need no explanation: the so-called natural log of any number is simply the power to which the Euler constant, *e*, which we saw earlier has a value of 2.7183, must be raised to equal the number. As *e* raised to the power -6.803 equals 0.00111, the natural log of 0.00111 (the rate constant measured at 650 K) is -6.803:

$$2.7183^{(-6.803)} = 0.00111$$

Pressing the ln key on your calculator, then entering 0.00111, takes care of this for you.

1	2	3	4	5
T	**1/T**	**1000/T**	**k**	**ln k**
(K)	**(K^{-1})**	**(K^{-1})**	**($dm^3\,mol^{-1}\,s^{-1}$)**	
650	0.00154	1.54	0.00111	-6.80
695	0.00144	1.44	0.00552	-5.20
745	0.00134	1.34	0.0612	-2.70
800	0.00125	1.25	0.674	-0.395
870	0.00115	1.15	4.06	1.40
950	0.00105	1.05	73.7	4.30
1030	0.000971	0.971	270	5.60
1080	0.000926	0.926	881	6.78

Table 2.3   Processing the rate constants for the second-order decomposition of ethanal to methane and carbon monoxide, measured over a range of temperatures, in preparation for the Arrhenius plot shown in **Figure 2.11** (see text for details).

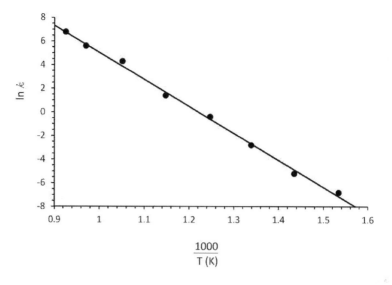

Figure 2.11   Logarithmic Arrhenius plot for the decomposition of ethanal to methane and carbon monoxide, using data from *columns 2* and *5* in **Table 2.3**.

Plotting the natural log of the rate constant (*column 5*) against '1000 divided by the temperature' (*column 3*) gives the straight-line graph shown in **Figure 2.11**. The simplest method for obtaining the gradient is to extend the line-of-best-fit such that it intercepts both axes, reading off these values to calculate the change in ln k for a given change in 1000/T. Thus, we see that increasing 1000/T from 0.90 to 1.58 (a change of 0.68) is accompanied by a change in ln k of −15.4 (from 7.4 to − 8.0). When working out the gradient, however, we must use the change in 1/T, rather than 1000/T (remember how the values of 1/T where multiplied by 1000 before plotting). The change 1/T, then, is 0.68 divided by 1000 (0.00068 K^{-1}) and the gradient is calculated as follows:[pp]

$$\text{gradient} \quad = \quad \frac{\text{change in ln } k}{\text{change in } 1/T}$$

$$= \quad \frac{-15.4}{0.00068 \ K^{-1}} \quad = \quad -22647 \ K$$

[pp] This is simply the perhaps more familiar Δy 'over' Δx, where y is ln k and x is 1/T. Logarithms are not given units, so the gradient has units of K.

As explained above, the gradient of the line-of-best-fit (m) corresponds to the $- E_a / R$ term in the logarithmic form of the Arrhenius equation, enabling us to determine the activation energy:

$$\text{gradient} \quad = \quad - \frac{E_a}{R}$$

$$- 22647 \text{ K} \quad = \quad - \frac{E_a}{8.314 \text{ J K}^{-1} \text{mol}^{-1}}$$

$$E_a \quad = \quad 22674 \text{ K} \times 8.341 \text{ J K}^{-1} \text{mol}^{-1}$$

$$= \quad 188287 \text{ J mol}^{-1}$$

$$= \quad 188 \text{ kJ mol}^{-1}$$

Notice how the minus signs on each side of the equation have been cancelled out (in effect, by multiplying both sides by −1) and how the K and K^{-1} unit cancels out, leaving the correct units of (k)J mol^{-1}. It now remains for us to obtain the Arrhenius constant, A.

When $1/T$ is zero, the natural log of A equals the natural log of k:

$$\ln k \quad = \quad - \frac{E_a}{R} \times \left(\frac{1}{T} \right) \quad + \quad \ln A$$

$$\ln k \quad = \quad - \frac{E_a}{R} \times 0 \quad + \quad \ln A$$

$$\ln k \quad = \quad \ln A$$

There are two ways to find the value of ln k when $1/T$ is zero. In the first method, ln k is simply read off the y-axis at the point of its intersection by the line-of-best-fit when $1/T$ is zero. This is not possible using the plot shown in **Figure 2.11**, of course, because the x-axis $(1/T)$ does not extend back to zero. In **Figure 2.12**, the data has been

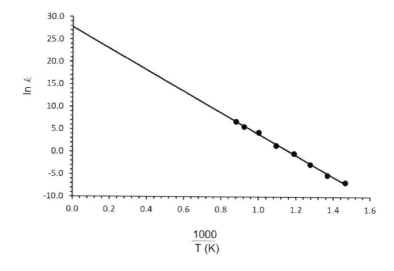

Figure 2.12 'Extended' Arrhenius plot for the decomposition of ethanal to methane and carbon monoxide, using data from *columns* 3 and 5 in **Table 2.3**. The line-of-best-fit is seen to intercept the *y*-axis (when 1/T equals zero) when ln *k* at 27.75.

replotted, but with the *x*-axis starting at zero, enabling ln *k* to be obtained when 1/T is zero, which is 27.75. We can say, therefore, that the natural log of the Arrhenius constant is 27.75:

$$\ln A \ = \ 27.75$$

The Arrhenius constant is now obtained by raising the Euler constant to the power 27.75 (using the $e^{\blacksquare}$ key on your calculator):

$$A \ = \ e^{27.75}$$

$$= \ 1.1(3) \times 10^{12} \ dm^3 \ mol^{-1} \ s^{-1}$$

Notice how – after 'coming back out' of logs – the appropriate units have been added to the Arrhenius constant; these are the units of the rate constant. In examination questions, you may be given a ready-plotted graph of the type shown in **Figure 2.12** and asked to determine the Arrhenius constant. On the other hand, you may be dealing with a graph that does not extend back to zero on the *x*-axis (**Figure 2.11**), in which case the Arrhenius constant can be obtained using the second method, where the Arrhenius equation is solved for A using a pair of data values and the activation energy you have calculated from the gradient. Substituting, for example, the first pair of values from **Table 2.3** (T = 650 K and k = 0.00111 dm^3 mol^{-1} s^{-1}) into the logarithmic form of the Arrhenius equation gives:

$$\ln 0.00111 \quad = \quad \left(-\frac{188287}{8.314} \times \frac{1}{650} \right) \quad + \quad \ln A$$

$$-6.80 \quad = \quad -34.8 + \ln A$$

$$28.0 \quad = \quad \ln A$$

$$A \quad = \quad e^{28.0}$$

$$= \quad 1.4 \times 10^{12} \, dm^3 \, mol^{-1} \, s^{-1}$$

Using this method, however, the value of the Arrhenius constant obtained is wholly dependent on the particular pair of temperature and rate constant values selected; if you were to use a different pair of values, you may well arrive at a different value for A. The advantage of using the first method is that, in using the line-of-best-fit, it takes into account every data point (every pair of temperature and rate constant values), thereby arriving at the 'best averaged' value for A.

We can achieve a similar outcome using the second method if, rather than choosing a pair of values from **Table 3.2**, we take a point directly on the line-of-best-fit. We have already noted that the line-of-best-fit in **Figure 2.11** crosses the y-axis where ln k equals 7.4 and 1/T is 0.00090 (1000/T is 0.90). Using this pair of values, which takes into account all the data points, we obtain:

$$7.4 \quad = \quad \left(-\frac{188287}{8.314} \times 0.00090 \right) + \ln A$$

$$27.8 \quad = \quad \ln A$$

$$A \quad = \quad 1.2 \times 10^{12} \, dm^3 \, mol^{-1} \, s^{-1}$$

CHAPTER 2 SUMMARY – the key points

- It can be helpful to think of a rate constant, k, as the rate of a given reaction observed when the reactants are present at a concentration of 1.00 mol dm^{-3}.

- For a reaction to take place, reactants must collide with sufficient energy to form the **transition state**, which is the point of highest energy in the reaction pathway.

- The energy changes during the course of a reaction are presented in a **reaction profile** graph.

- The amount of energy with which reactant particles must collide to form the transition state is the **activation energy** of the reaction, E_a.

- Boltzmann distributions give the distribution of energy within populations of particles at a stated temperature. In the case of an ideal gas, the **Boltzmann distribution** shows how the total kinetic energy of the body of gas is distributed between its individual gaseous molecules. The distribution of kinetic energy within a population of gaseous particles at a stated temperature determines the fraction of particles that are able to collide with a specified minimum energy.

- The frequency of all the collisions – irrespective of the energy with which they occur – within a population of reactant particles, in an orientation that can result in a chemical reaction, is given by the **Arrhenius constant**, A. The proportion (or fraction) of the **collisions** that, at a stated temperature, occur with an energy of at least the activation energy – irrespective of their frequency and geometry – is given by $e^{-(E_a/RT)}$.

- The rate constant of a reaction at a given temperature is equal to the frequency of the collisions that occur with an energy of at least the activation energy and in a geometrical arrangement that allows the transition state to be formed. It is calculated using the **Arrhenius equation**, in which A and $e^{-(E_a/RT)}$ are multiplied together:

$$k = A\,e^{-(E_a/RT)}$$

- The logarithmic form of the Arrhenius equation,

$$\ln k = -E_a/RT + \ln A$$

is used to obtain the activation energy and Arrhenius constant from experimental data. Plotting $\ln k$ on the y-axis against $1/T$ on the x-axis gives a straight line, the gradient of which equals $-E_a/R$. The point at which the line intercepts the y-axis is equal to $\ln A$.

Chapter 3

DERIVING REACTION MECHANISMS FROM RATE EQUATIONS: 'PUTTING THE HORSE BEFORE THE CART'

The preceding two chapters have already provided considerable insight into how the mechanism of a given chemical reaction – the 'route' by which its reactants are turned into products – is reflected in its rate equation. Thus, in Chapter 1 we saw that only the reactants involved in the slowest, rate-determining step appear in the rate equation; and in Chapter 2 we saw that the magnitude of the rate constant is determined primarily by the activation energy of the reaction: the higher the activation energy, the smaller the rate constant. To illustrate the relationship between activation energies and rate constants, two relatively 'uncomplicated' reactions were selected as examples – reactions that take place in a single step, namely the nucleophilic substitution reaction between the hydroxide ion and bromoethane,

$$\text{:}\overset{-}{\text{O}}\text{H} \quad + \quad \overset{H}{\underset{H_3C}{\overset{|}{C}}}{-}Br \quad \longrightarrow \quad \left[HO\text{----}\overset{H}{\underset{H \quad CH_3}{C}}\text{----}Br \right]^{-} \quad \longrightarrow \quad HO{-}\overset{H}{\underset{CH_3}{C}}{\text{'''}}H \quad + \quad \text{:}\overset{-}{Br}$$

transition state

and the hydrogenation of ethene:

$$\overset{H}{\underset{H}{>}}C{=}C\overset{H}{\underset{H}{<}} \quad + \quad H{-}H \quad \longrightarrow \quad \left[\overset{H \quad H}{\underset{H\text{----}H}{H{-}\overset{|}{C}{-}\overset{|}{C}{-}H}} \right] \quad \longrightarrow \quad \overset{H \quad H}{\underset{H \quad H}{H{-}\overset{|}{C}{-}\overset{|}{C}{-}H}}$$

transition state

In both these second-order reactions, the energy required to generate the transition state – in a direct collision between the two reactants – corresponds to the activation energy, which, along with the Arrhenius constant (A), determines the magnitude of the rate constant. In our detailed exploration of how these factors are linked through the Arrhenius equation, the discussion was confined to such single-step reaction mechanisms. In Chapter 1, however, we encountered reactions that occur in multiple steps, these being the nucleophilic substitution reaction of the hydroxide ion with a *tertiary* halogenoalkane, 2-bromomethylpropane, and the oxidation of the bromide ion by the bromate(V) ion.

Each of the individual steps in a multistep reaction mechanism involves a collision; it follows, therefore, that each step will have its own activation energy. It is, however, only the step with the largest activation energy that is reflected in the rate equation and determines the magnitude of the rate constant for the overall reaction. *The rate constant of a multistep reaction is, in fact, the rate constant for its slowest step.* These ideas will now be developed further using the examples discussed below.

3.1   Reactions of halogenoalkanes

It was described in Chapter 1 how the first-order nucleophilic substitution reaction between the hydroxide ion and 2-bromomethylpropane – a tertiary halogenoalkane – involves the initial formation of a tertiary carbocation in a slow step (the rate-determining step), followed by the relatively rapid addition of OH⁻ to the carbocation:

We saw at the close of the chapter that the rate equation for this reaction is simply,

$$\text{rate} = k\,[(CH_3)_3CBr]$$

reflecting the finding that the rate of the reaction is insensitive to the concentration of hydroxide ions.

Notice that, whereas both 2-bromomethylpropane and the product, methylpropan-2-ol, are tetrahedral about the carbon atom in position 2, the carbocation is trigonal planar.[qq] This is because there are only three pairs of electrons around the carbon bearing the positive charge. In the reaction scheme above, the hydroxide ion is shown approaching and adding to the carbocation from above. There is, however, an equal probability of OH^- approaching and adding from below the plane of the carbocation, which would result in the geometry about the central carbon in the product being inverted relative to that shown here. As there are three methyl groups attached to the central carbon in the carbocation seen in this example, the product will be identical, irrespective of whether OH^- approaches from above or below the plane. If, however, there were three different alkyl (or aryl) groups attached directly to the central carbon atom, a 50:50 mixture of enantiomers – a *racemic mixture* – would be formed, reflecting the equal probability of OH^- approaching the carbocation from above and below.[rr]

The topic of optical isomerism is outside the scope of this book, but it is covered in the main A-level chemistry textbooks, where you will find similar discussions on the generation of racemic mixtures during the nucleophilic addition of the cyanide ion to carbonyl compounds. Recall, also, how the geometry around the carbon at which substitution takes place is inverted in second-order nucleophilic substitution reactions, such as that between the hydroxide ion and 1-bromoethane (see page 14).

[qq] For an explanation of the shapes of molecules and ions (based on electron-pair repulsion theory), see the YouTube page, 'Westcott Research and Consulting'.

[rr] Secondary halogenoalkanes undergo nucleophilic substitution by both the first- and second-order mechanisms, the preferred route depending on the reaction conditions. The carbocation formed from a secondary halogenoalkane would have a hydrogen atom in place of one of the alkyl (or aryl) groups, which would still allow for the formation of a racemic mixture.

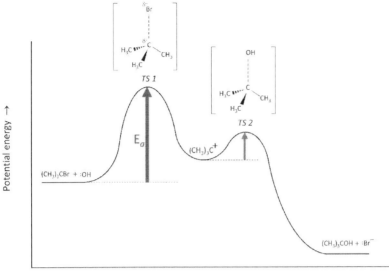

Figure 3.1 Reaction profile for the reaction between the hydroxide ion and 2-bromomethylpropane. An OH⁻ ion and a Br⁻ ion are also present with the carbocation in the energy 'well', but these have been omitted for clarity. The arrows show the activation energy associated with the formation of each of the two transition states, *TS 1* and *TS 2*. The larger arrow, labelled E_a, is the activation energy for the rate-determining step and, therefore, determines the rate constant for the overall reaction.

The reaction profile for the nucleophilic substitution reaction between the hydroxide ion and 2-bromomethylpropane is shown in **Figure 3.1**. Notice the important difference between a transition state and a reaction intermediate (the carbocation). Whereas transition states occur at potential energy maxima (on the 'peaks'), reaction intermediates occur at the minima (in the 'wells'). Two transition states are shown: the first (*TS 1*) occurs when the carbon-to-bromine bond has been stretched to its maximum length before breaking; in the second transition state (*TS 2*), the hydroxide ion is just close enough to the carbocation for the new bond (oxygen-to-carbon) to begin forming. There is an activation energy associated with the formation of each transition state, shown by the two arrows. Notice, however, that the activation energy for the formation of *TS 1* is much larger than that associated with the formation of *TS 2*. This is the reason why the formation of the carbocation is the rate-determining step and also why, once formed, the carbocation does not 'hang around': the energy needed for the carbocation to 'disappear' by reacting with the hydroxide ion is very small compared with the energy required for its formation from 2-bromomethylpropane. This key concept can be stated thus:

> **In any reaction that involves multiple steps (and therefore multiple transition states), the step with the highest activation energy is the rate-determining step.**

It follows, therefore, that *when we are measuring the rate of a reaction that occurs in more than one step, we are, in fact, measuring only the rate of its slowest step.* In the above example, we are measuring the rate of formation of the carbocation: the carbocation is converted to methylpropan-2-ol so rapidly, there is effectively no difference between the rate of formation of the carbocation and that of the alcohol. Changing the concentration of the hydroxide ion, which has no role in the rate-determining step, has no effect on the rate of this reaction.

You may be surprised to learn, then, that reducing the concentration of the OH^- ion even to zero has no effect on the rate of methylpropan-2-ol formation. This is because H_2O can 'take the place' of OH^- as the nucleophile in the reaction. The water molecule has all the attributes of a nucleophile: it has a lone pair of electrons (two, in fact) to form a covalent bond to the positively-charged carbon of the carbocation; and the oxygen atom carrying the required lone pair has a partial-negative charge.[ss]

[ss] Just as the negative charge on a nucleophile can be either a full charge (−) or a partial charge (δ−), the positive charge on the carbon to which it is attracted and forms a new covalent bond can be either a full charge (+) or partial charge (δ+). Consider, for example, the nucleophilic substitution reaction between ammonia and 1-bromoethane, where the nucleophilic N atom in NH_3 carries a partial-negative charge.

A simple way to understand how water behaves as a nucleophile is to view the H_2O molecule as an OH^- ion with an 'added' H^+ ion:

$$OH^- + H^+ \rightarrow H_2O$$

The H_2O molecule simply releases the 'extra' proton – **deprotonates** – *after* it has added to the positively-charged carbon:

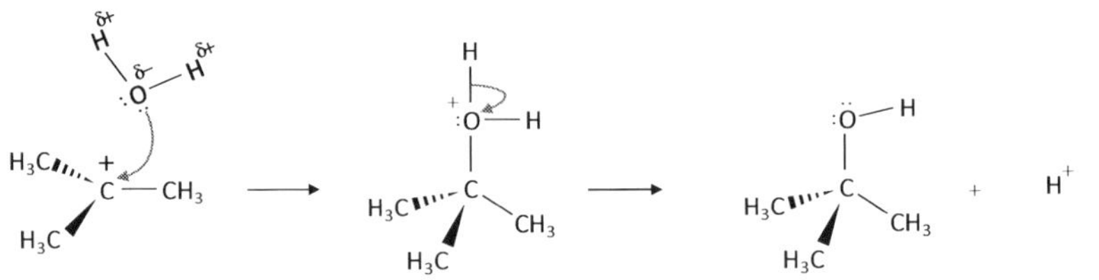

As you will have no doubt anticipated, there is another transition state associated with the deprotonation step, but, as with the addition of H_2O to the carbocation, the activation energy for this step is very small compared with that for the formation of the carbocation. In other words, the deprotonation step is so fast, it does not affect the rate of the overall reaction, which is determined only by the rate at which the carbocation is formed.

The H_2O molecule, of course, lacks a *full* negative charge on its oxygen atom, making it a less powerful nucleophile than the OH^- ion – that is to say, it is attracted less powerfully than OH^- to the carbocation. Although replacing OH^- by H_2O as the nucleophile, then, slows this individual step of the reaction, it has **no effect on the overall rate of a first-order substitution reaction**, which – *I will repeat only one more time!* – is determined only by the rate at which the carbocation is formed.

Exchanging H_2O for OH^- in a *second*-order nucleophilic substitution, however, **does** result in a lower rate of reaction. This is because such reactions involve a direct collision between the nucleophile and the halogenoalkane: the nucleophile does not have to 'wait' for a carbocation to be formed. Since the carbon atom in the halogenoalkane with a partial-positive charge ($\delta+$) attracts H_2O less powerfully than it attracts the OH^- ion, the reaction is slower:

Unlike the situation in a second-order nucleophilic substitution reaction, there is nothing to be gained by replacing water with OH⁻ as the nucleophile in a first-order reaction.

Indeed, adding alkali to a first-order nucleophilic substitution reaction would be counterproductive: it would, in fact, *lower* the yield of the alcohol. This is because, as well as being a good nucleophile, the OH⁻ ion is a powerful base. Students following the AQA and Edexcel specifications will know that halogenoalkanes also undergo elimination reactions with the hydroxide ion, as illustrated in the reaction involving the primary halogenoalkane bromoethane:

In an elimination reaction, the hydroxide ion removes a H^+ ion from a carbon atom *adjacent* to the carbon to which the halogen atom is attached – the carbon at position 2 in bromoethane. By accepting a proton, OH^- is reacting as a base, rather than as a nucleophile. When a halogenoalkane is attacked by the OH^- ion, there is always the possibly of both substitution and elimination occurring. With primary halogenoalkanes, the predominant reaction is nucleophilic substitution, so the elimination reaction shown above for bromoethane represents a minor reaction: the major product would be ethanol.

Tertiary halogenoalkanes, however, react with OH^- primarily by elimination, forming the corresponding alkene. This is because the extra alkyl groups in a tertiary halogenoalkane hinder the access of the approaching OH^- ion to the carbon atom at which substitution would take place: alkyl groups are bulky, so the OH^- encounters a H atom on an alkyl group before it ever gets close enough to the carbon for substitution to occur. The H_2O molecule is a much weaker base (H^+ acceptor) than the OH^- ion so, rather than taking a proton off one of the alkyl groups, it perseveres on its way to the carbon, reacting as a nucleophile. This is why replacing OH^- with H_2O in the reaction with 2-bromomethylpropane reduces the yield of the alcohol, methylpropan-2-ol, producing, instead, the corresponding alkene *via* elimination:

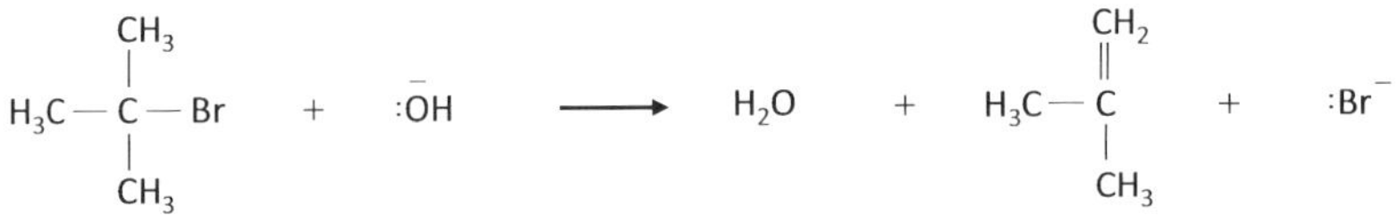

When a tertiary halogenoalkane, such as 2-bromomethylpropane, undergoes an elimination reaction with the hydroxide ion, the mechanism is somewhat different to that shown above for the reaction of a primary halogenoalkane (bromoethane): tertiary halogenoalkanes initially form a carbocation, from which the proton is removed. As you might expect, the formation of the carbocation is the rate-determining step and the reaction displays first-order kinetics.[tt]

[tt] The mechanism of first-order elimination is not covered at A-level, but always check the latest version of your specification.

As a general rule, then, primary halogenoalkanes react with the hydroxide ion predominantly by substitution and tertiary halogenoalkanes by elimination; secondary halogenoalkanes react by a mixture of the two. We can, however, change the conditions to favour either route for a given halogenoalkane: substitution is favoured by using dilute, aqueous alkali[uu] at room temperature; elimination is favoured by using hot, concentrated, ethanolic alkali. We can explain how the conditions affect the type of reaction that occurs by considering the effects of the concentration of the alkali, the solvent and temperature individually.

The hydroxide ion is a powerful base, so to minimise the elimination reaction, in which it is acting as a base, the ion is used at low concentration. Although by keeping the concentration of OH^- low we are also limiting the availability of the ion for participation in the substitution reaction, there is an abundance of H_2O molecules from the solvent to also act as the nucleophile. Consider, for example, the substitution reactions that take place when 2-bromopropane, a secondary halogenoalkane, reacts with either the OH^- ion or the H_2O molecule, to give propan-2-ol:

$$CH_3CHBrCH_3 \ + \ OH^- \ \rightarrow \ CH_3CH(OH)CH_3 \ + \ Br^-$$

$$CH_3CHBrCH_3 \ + \ H_2O \ \rightarrow \ CH_3CH(OH)CH_3 \ + \ HBr$$

When carried out in a 60 % solution of ethanol in water (by volume), at a temperature of 55 °C, both reactions are found to be *second* order: they involve the direct attack of the nucleophile at the δ^+-carbon atom in position 2 of the halogenoalkane – essentially, as shown on page 97 for the corresponding reactions of bromoethane with OH^- and H_2O (but where attack occurs at carbon 1). The respective rate equations are,

$$\text{rate} \ = \ k_{(OH^-)} \, [OH^-][CH_3CHBrCH_3]$$

$$\text{rate} \ = \ k_{(H_2O)} \, [H_2O][CH_3CHBrCH_3]$$

[uu] Halogenoalkanes are not soluble in water and are, therefore, usually dissolved in ethanol before adding the aqueous alkali. Thus, although the alkali solution is entirely aqueous, the final reaction mixture will always contain *some* ethanol.

where $k_{(OH^-)}$ and $k_{(H_2O)}$ are the corresponding second-order rate constants, with values of 3.0×10^{-5} and 3.0×10^{-7} dm^3 mol^{-1} s^{-1}, respectively.

Although the rate constant for the reaction with water is only one-hundredth of that with the hydroxide ion (H_2O is a much weaker nucleophile than the OH^- ion[vv]), this is compensated for when the concentration of H_2O is much higher than that of OH^- ion. This is illustrated by the following example, in which we first calculate the rate of the substitution reaction between 2-bromopropane, at a concentration of 0.20 mol dm^{-3}, and OH^- in dilute alkali (0.10 mol dm^{-3} KOH), with the reaction taking place in 60 % ethanol at 55 °C:

$$\text{rate} = k_{(OH^-)}[OH^-][CH_3CHBrCH_3]$$

$$= (3.0 \times 10^{-5}\, dm^3\, mol^{-1}\, s^{-1})(0.10\, mol\, dm^{-3})(0.20\, mol\, dm^{-3})$$

$$= 6.0 \times 10^{-7}\, mol\, dm^{-3}\, s^{-1}$$

In this solvent mixture (60 % ethanol, 40 % water), the concentration of H_2O molecules is 22.2 mol dm^{-3},[ww] from which we can calculate the rate of the corresponding reaction involving H_2O as the nucleophile:

$$\text{rate} = k_{(H_2O)}[H_2O][CH_3CHBrCH_3]$$

$$= (3.0 \times 10^{-7}\, dm^3\, mol^{-1}\, s^{-1})(22.2\, mol\, dm^{-3})(0.20\, mol\, dm^{-3})$$

$$= 1.3 \times 10^{-6}\, mol\, dm^{-3}\, s^{-1}$$

It is seen that, in *dilute* alkali, the reaction of H_2O with the halogenoalkane makes a greater contribution, than the corresponding reaction involving the OH^- ion, to the rate at which propan-2-ol is formed. This means that, by using dilute alkali, we can minimise the rate of propene generation in the elimination reaction, yet still form the alcohol. The combined, *total* rate at which propan-2-ol is formed under

---

[vv] As the oxygen atom in water has only a partial negative charge (δ^-), it is not attracted as powerfully as the OH^- ion to the δ^+-carbon atom at position 2 in $CH_3CH_2BrCH_3$.

[ww] The mass of 1 dm^3 of water (1000 cm^3) is 1000 g. Since 1000 g of water contains 55.56 moles of H_2O molecules (1000 g/18 g mol^{-1}), the concentration of H_2O in pure water is 55.56 mol dm^{-3} and that of a 60:40 ethanol/water mixture (by volume), therefore, 22.2 mol dm^{-3}.

these conditions – involving substitution by both H_2O and the OH^- ion – is obtained by adding together the two individual rates (1.3×10^{-6} and 6.0×10^{-7} mol dm^{-3} s^{-1}) and amounts to 1.9×10^{-6} mol dm^{-3} s^{-1}.

The equation for the formation of propene from 2-bromopropane in the elimination reaction is:

$$CH_3CHBrCH_3 \; + \; OH^- \; \rightarrow \; CH_3CHCH_2 \; + \; H_2O \; + \; Br^-$$

This reaction is also second order, having a rate constant, $k_{(elimination)}$, of 4.7×10^{-5} dm^3 mol^{-1} s^{-1} in a 60 % solution of ethanol in water at 55 °C. When we calculate the rate of this reaction under the conditions used above for the corresponding substitution reaction, it is seen to proceed a rate that is about half that of the combined substitution reaction:

$$\text{rate} \; = \; k_{(elimination)} \, [OH^-][(CH_3CHBrCH_3]$$

$$= \; (4.7 \times 10^{-5} \, \text{dm}^3 \, \text{mol}^{-1} \, \text{s}^{-1})(0.10 \, \text{mol dm}^{-3})(0.20 \, \text{mol dm}^{-3})$$

$$= \; 9.4 \times 10^{-7} \, \text{mol dm}^{-3} \, \text{s}^{-1}$$

If we now carry out the same calculations but using *concentrated* alkali (2.0 mol dm^{-3} KOH), it is seen that substitution by OH^- ion occurs at a rate of 1.2×10^{-5} mol dm^{-3} s^{-1}:

$$\text{rate} \; = \; k_{(OH^-)} \, [OH^-][CH_3CHBrCH_3]$$

$$= \; (3.0 \times 10^{-5} \, \text{dm}^3 \, \text{mol}^{-1} \, \text{s}^{-1})(2.0 \, \text{mol dm}^{-3})(0.20 \, \text{mol dm}^{-3})$$

$$= \; 1.2 \times 10^{-5} \, \text{mol dm}^{-3} \, \text{s}^{-1}$$

Since the concentration of water is still 22.2 mol dm^{-3}, the rate of substitution with H_2O as the nucleophile is still 1.3×10^{-6} mol dm^{-3} s^{-1}, giving a combined rate for the two substitution reactions of 1.3×10^{-5} mol dm^{-3} s^{-1}. Under the same conditions, elimination occurs a rate of 1.9×10^{-5} mol dm^{-3} s^{-1}, which is now *greater* than the combined rate for the substitution reaction:

$$\text{rate} \; = \; k_{(elimination)} \, [OH^-][CH_3CHBrCH_3]$$

$$= \; (4.7 \times 10^{-5} \, \text{dm}^3 \, \text{mol}^{-1} \, \text{s}^{-1})(2.0 \, \text{mol dm}^{-3})(\, 0.20 \, \text{mol dm}^{-3})$$

$$= \; 1.9 \times 10^{-5} \, \text{mol dm}^{-3} \, \text{s}^{-1}$$

Thus, by increasing the concentration of alkali from 0.1 to 2.0 mol dm^{-3}, the balance of the two reaction pathways has been shifted from substitution to elimination: the rate of elimination has increased from being half to 1.5-times the rate of elimination. (This is because [OH$^-$] has been increased, but not [H$_2$O], so there has been no increase in the rate of substitution by H$_2$O to balance the increase in the rate of elimination by OH$^-$.)

The composition of the solvent in which these reactions take place – namely, its ethanol-to-water ratio – also has a direct effect on the relative rates of the two pathways. The rate constants for both the nucleophilic substitution and elimination reactions are very sensitive to the nature of the solvent. A knowledge of the reasons for this sensitivity is not required at A-level, so will not be covered here: suffice to say, they involve the polarity of the solvent, which affects the ease with which different transition states are formed, thereby determining activation energies and hence rate constants. In 100 % ethanol, then, the rate constants for substitution and elimination by the OH$^-$ ion at 55 °C are, respectively:

$$k_{(OH^-)} = 6.0 \times 10^{-5} \, dm^3 \, mol^{-1} \, s^{-1}$$

$$k_{(elimination)} = 1.5 \times 10^{-4} \, dm^3 \, mol^{-1} \, s^{-1}$$

Notice how $k_{(elimination)}$ is now 2.5-times greater than $k_{(OH^-)}$. In 60 % ethanol, the difference was only 1.6-fold. It follows then that, whatever the concentrations of CH$_3$CHBrCH$_3$ and OH$^-$ used, the rate of elimination reaction will always be 2.5-times faster than the rate of substitution.[xx]

The effects we have seen – of increasing the concentration of the alkali and replacing water with ethanol as the solvent – are not dramatic, but clearly favour elimination over substitution. This brings us to the effect of temperature. As we saw in Chapter 2, the greater the activation energy (E_a) of a reaction, the greater its sensitivity to temperature changes. **Table 3.1** gives the Arrhenius parameters, E_a and the Arrhenius constant (A), for the reactions of 2-bromopropane with water (by nucleophilic substitution) and the hydroxide ion (by nucleophilic substitution and elimination) in 80 % ethanol – *some* water must be

---

[xx] In 100 % ethanol, there are of course no H$_2$O molecules to contribute to the substitution reaction.

included to react with the halogenoalkane. As mentioned above, rate constants are sensitive to the nature of the solvent, which involves its effects on both E_a and A, therefore the rate constants we will calculate from these values apply only to this solvent mixture.

Reaction	A / dm^3 mol^{-1} s^{-1}	E_a /J mol^{-1}
$CH_3CHBrCH_3$ + KOH $\rightarrow$ CH_3CHCH_2 + H_2O + KBr (Elimination by OH$^-$)	8.38×10^{10}	94726
$CH_3CHBrCH_3$ + KOH $\rightarrow$ $CH_3CH(OH)CH_3$ + KBr (Nucleophilic substitution by OH$^-$)	1.40×10^{10}	90793
$CH_3CHBrCH_3$ + H_2O $\rightarrow$ $CH_3CH(OH)CH_3$ + HBr (Nucleophilic substitution by H_2O)	6.94×10^{9}	96901

Table 3.1   Arrhenius parameters for the reactions of the hydroxide ion and water with 2-bromopropane in an 80 % (by volume) solution of ethanol in water.

The activation energy for the elimination reaction between the OH$^-$ ion and $CH_3CHBrCH_3$ is higher than that for the substitution reaction, which means that increasing the temperature will shift the balance of the two processes towards elimination. Although the substitution reaction with H_2O as the nucleophile has an even higher activation energy (and so will be increasingly favoured at the expense of elimination as the temperature is increased), this reaction can be prevented altogether by reacting OH^- with $CH_3CHBrCH_3$ in pure ethanol. Notice also that the Arrhenius constant, A, is much higher for the elimination reaction with OH^- than for the corresponding substitution reaction. Recall that the Arrhenius constant, A, is also referred to as the collision frequency factor: it gives the frequency of collisions that occur *in an orientation that allows the transition state to be formed.* In a secondary halogenoalkane, such as $CH_3CHBrCH_3$, the two bulky alkyl groups will restrict the access of the OH^- ion to the carbon atom to which the Br is bonded, resulting in a relatively low collision frequency factor. In contrast, there are six possible H atoms that can be removed as H^+ ions when OH^- collides with $CH_3CHBrCH_3$ in the elimination reaction. As described on page 48, the Arrhenius constant is really two constants 'rolled into one' – the collision frequency factor, Z, and the steric factor, P. It is the latter component, P, that is responsible for the relatively low value of A in the substitution reaction.

In **Table 3.2**, the rate constants for all three reactions – calculated using the Arrhenius parameters given in **Table 3.1** – are given at two temperatures. Even though it has a higher activation energy, the rate constant for the elimination reaction between OH^- and $CH_3CHBrCH_3$ is higher than that for substitution at both temperatures. This is a reflection of the contribution made by the smaller steric factor, P, to the Arrhenius constant (and hence rate constant) for the substitution reaction. The factor by which elimination is favoured over substitution is seen to increase at higher temperature. Thus, at 25 °C the rate constant for the elimination reaction is 1.2-times that of the substitution reaction with OH^-, whereas at 90 °C the rate constant for elimination is 1.6-times greater. Although the rate constant for substitution with H_2O acting as the nucleophile is much lower at both temperatures, the concentration of water in 80 % ethanol is still high enough (11.1 mol dm^{-3}) for this reaction to make a significant contribution to the formation of the alcohol. Substitution with H_2O as the nucleophile can, however, be avoided entirely by carrying out the reaction in 100 % ethanol, which – through its effects on E_a and A – will also affect the rate constants for substitution and elimination by the OH^- ion, favouring the latter. In 100 % ethanol at, for example, 55 °C, the rate constant for elimination is 2.4-times greater than that for substitution (1.46 × 10^{-4} and 0.60 × 10^{-4} dm^3 mol^{-1} s^{-1}, respectively).

Reaction	$k\,/\mathrm{dm^3\,mol^{-1}\,s^{-1}}$	
	25 °C	90 °C
$CH_3CHBrCH_3 + KOH \rightarrow$ $CH_3CHCH_2 + H_2O + KBr$ (Elimination by OH^-)	2.08×10^{-6}	1.96×10^{-3}
$CH_3CHBrCH_3 + KOH \rightarrow$ $CH_3CH(OH)CH_3 + KBr$ (Nucleophilic substitution by OH^-)	1.70×10^{-6}	1.20×10^{-3}
$CH_3CHBrCH_3 + H_2O \rightarrow$ $CH_3CH(OH)CH_3 + HBr$ (Nucleophilic substitution by H_2O)	7.17×10^{-8}	7.89×10^{-5}

Table 3.2   Calculated rate constants for the reactions of the hydroxide ion and water with 2-bromopropane at two temperatures in an 80 % (by volume) solution of ethanol in water. See text for details.

Although there is no set of conditions we can select to ensure the OH^- ion reacts with a secondary halogenoalkane, such as 2-bromopropane, *solely* by elimination or substitution, we can shift the balance to favour one pathway over the other through our choice of temperature, solvent and alkali concentration. Also bear in mind that, primary halogenoalkanes react with OH^- predominantly by substitution and tertiary halogenoalkanes almost exclusively by elimination, making it much easier to favour a single reaction pathway for such molecules.

3.2 Acid-catalysed halogenation of propanone

Halogens react with propanone and most other carbonyl compounds in an acid-catalysed reaction, as shown below for iodine:[yy]

$$CH_3COCH_3 + I_2 \rightarrow CH_3COCH_2I + HI$$

It has been established that this reaction is zero-order with respect to iodine. In other words, its rate is not affected by changes in the concentration of iodine. The reaction is, however, first-order with respect to the concentration of both propanone and the H^+ ion, making it second order overall. The rate equation is:

$$\text{rate} = k[H^+][CH_3COCH_3]$$

It is the job of the physical organic chemist to propose a reaction mechanism that is consistent with the experimentally-derived rate equation.

Ketones and aldehydes undergo reversible isomerisation to their corresponding enols,[zz] as shown here for propanone:

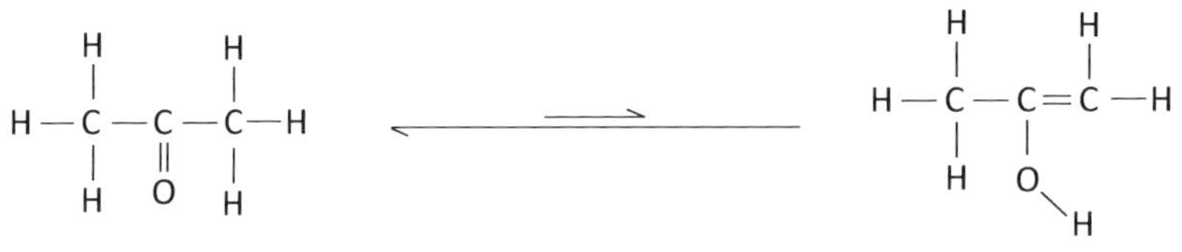

Ketone form
(propanone)

Enol form
(propene-2-ol)

Enols are much less stable than their corresponding ketones, therefore the above equilibrium lies far to the left (as indicated by the relative lengths of the reaction arrows).

[yy] Aldehydes and ketones with a hydrogen atom bonded to a carbon atom *adjacent* to the carbonyl group undergo this reaction: as shown in the equation for the reaction of iodine with propanone, the hydrogen atom in this position is replaced by a halogen atom.

[zz] The name 'enol' reflects the fact that these compounds are both alkenes and alcohols: 'en(e)' and 'ol'

The equilibrium constant, K_c,

$$K_c = \frac{[enol]}{[ketone]}$$

is 4.69×10^{-9} for propanone at 25 °C. Some textbooks report higher values, between 10^{-7} and 10^{-5}, but whichever value one takes, it is evident that the enol form of propanone is present at only trace levels. It is, however, in their enol forms that ketones and aldehydes react with halogens. The high electron-density in the carbon-to-carbon double bond of the enol induces a dipole in the halogen molecule, allowing it to attack the double bond as an electrophile:

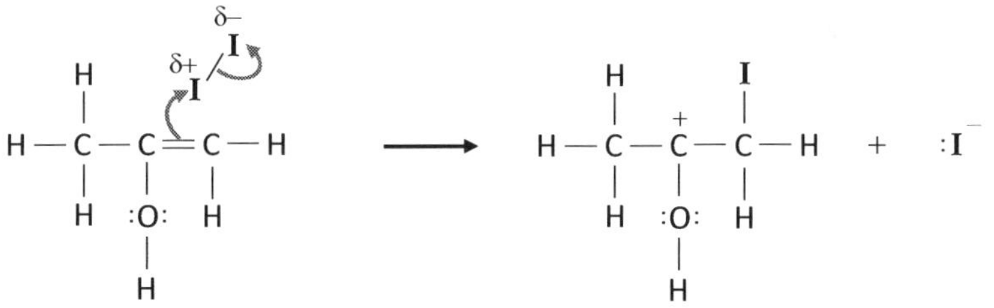

Rather than the iodide ion then adding to the carbocation (as would occur when a halogen adds to a simple alkene), the positive charge is now transferred to the oxygen atom, followed by the release of a H$^+$ ion (deprotonation):

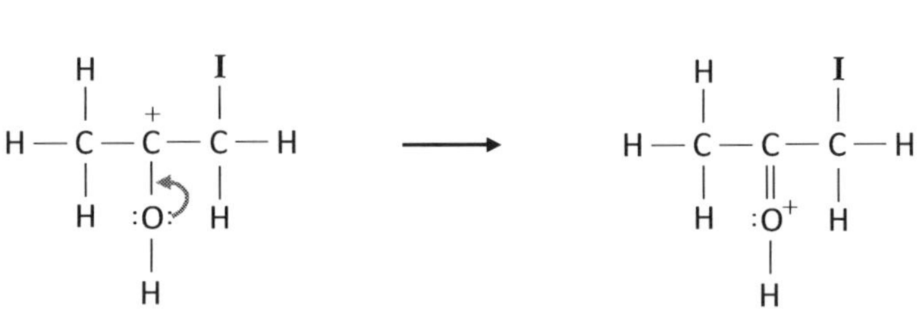

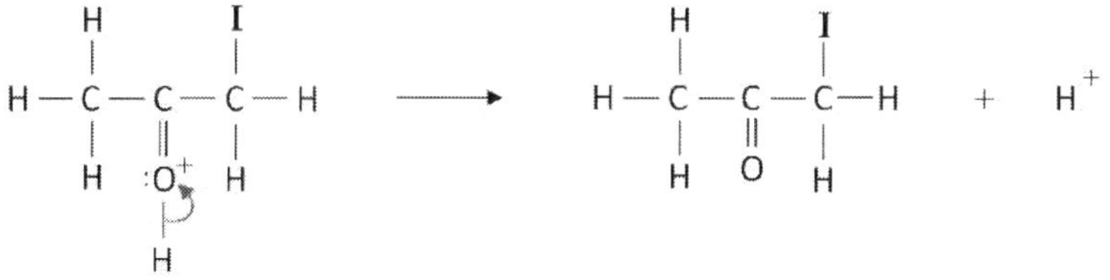

Iodine does not appear in the rate equation,

$$\text{rate} \quad = \quad k\,[H^+][CH_3COCH_3]$$

which tells us that the reaction of iodine with the enol form of propanone, through the mechanism shown above, cannot be the rate-determining step in the *overall* reaction. Instead, the rate equation indicates that the rate-determining step involves an interaction between the H^+ ion and propanone. This reflects the ability of H^+ ions to catalyse the conversion of ketones and aldehydes to their corresponding enols, shown here for propanone:

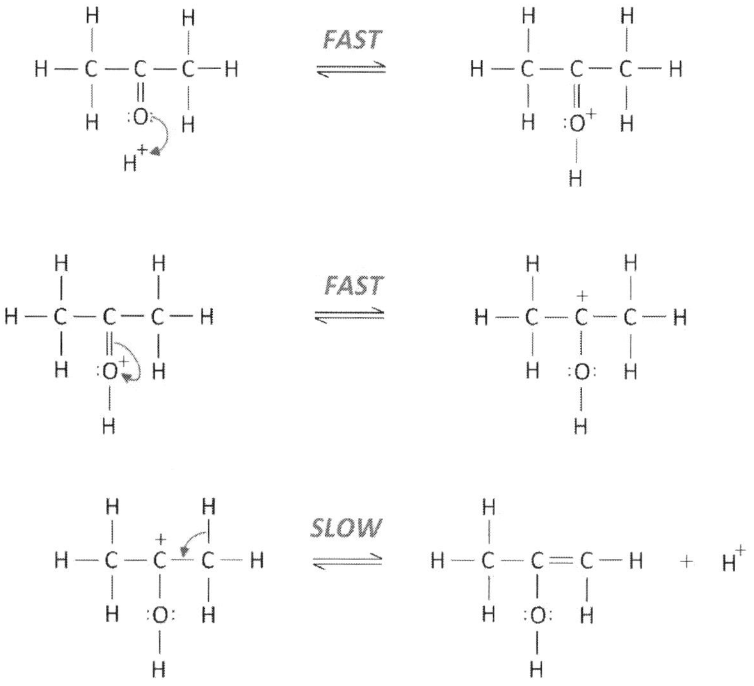

Notice how the H^+ ion that adds to propanone in the initial step is regenerated in the final step and is, thereby, acting as a catalyst. As indicated, the final step, in which a H^+ ion is released, is the slowest step in the formation of the enol. Moreover, this step is the slowest and, therefore, the rate-determining step, in the overall reaction between iodine and propanone, which we can summarise in two groups of elementary reactions:

Acid-catalysed generation of the enol

$$CH_3COCH_3 + H^+ \rightarrow CH_3C(O^+H)CH_3 \qquad \text{(fast)}$$

$$CH_3C(O^+H)CH_3 \rightarrow CH_3C^+(OH)CH_3 \qquad \text{(fast)}$$

$$CH_3C^+(OH)CH_3 \rightarrow CH_3C(OH)CH_2 + H^+ \quad \text{(slow)}$$

Reaction of iodine with the enol

$$CH_3C(OH)CH_2 + I_2 \rightarrow CH_3C^+(OH)CH_2I + I^- \quad \text{(fast)}$$

$$CH_3C^+(OH)CH_2I \rightarrow CH_3C(O^+H)CH_2I \qquad \text{(fast)}$$

$$CH_3C(O^+H)CH_2I \rightarrow CH_3COCH_2I + H^+ \qquad \text{(fast)}$$

These two groups of reaction steps may be combined into two overall steps, in which the formation of the enol – which cannot go any faster than its slowest step (deprotonation) – is the slowest, rate-determining step:

$$CH_3COCH_3 + H^+ \rightarrow CH_3C(OH)CH_2 + H^+ \qquad \text{(slow)}$$

$$CH_3C(OH)CH_2 + I_2 \rightarrow CH_3COCH_2I + H^+ + I^- \quad \text{(fast)}$$

Adding these together (and cancelling out the enol and the catalytic H^+ ion), gives the overall reaction:

$$CH_3COCH_3 + I_2 \rightarrow CH_3COCH_2I + H^+ + I^-$$

It is, however, only by considering the reaction to occur in at least two stages that we can see why it is zero order with respect to iodine: increasing the concentration of iodine is futile because the halogen

109

cannot participate in the reaction until the enol has been generated, which is 'holding up' the whole process. As we saw earlier when considering first-order nucleophilic substitution reactions (page 16), this scenario can be compared with making a cake that has a single cherry on top: adding the cherry is the fast step, so waiting by the oven with hundreds of cherries, ready to pop one on the cake when it comes out of the oven, is not going to speed up the overall process. The time it takes to make our cake is determined by the slowest step, during which it is baking in the oven.

In a reaction such as this, it is easy to confuse the concepts of rate and equilibrium. If we were to prepare a solution of propanone at a concentration of, let's say, 1.00 mol dm^{-3}, then, based on the equilibrium constant given above, the concentration of propene-2-ol would be $4.69 \times 10^{-9} \text{ mol dm}^{-3}$, which is tiny compared with that of the ketone:

$$CH_3COCH_3 \rightleftharpoons CH_3C(OH)CH_2$$

The addition of iodine, in excess, results in the rapid removal of the enol, forming iodopropanone:

$$CH_3C(OH)CH_2 + I_2 \rightarrow CH_3COCH_2I + HI$$

The rate constant for this reaction – the reaction of iodine with propanone *in its enol form* – is indeed very high, being $4.0 \times 10^9 \text{ dm}^3 \text{ mol}^{-1} \text{ s}^{-1}$. In order to restore the relative concentrations of the enol and ketone forms of propanone to those prescribed by the equilibrium constant,

$$K_c = \frac{[\text{propene-2-ol}]}{[\text{propanone}]} = 4.69 \times 10^{-9}$$

the enol removed in the reaction with iodine must be replaced by the isomerisation of more propanone. The rate equation for the isomerisation of propanone to propene-2-ol is,

$$\text{rate} = k_E[H^+][CH_3COCH_3]$$

where k_E, the second-order rate constant, is $2.79 \times 10^{-5} \text{ dm}^3 \text{ mol}^{-1} \text{ s}^{-1}$. (The subscripted E is to indicate that this is the rate constant for the reaction in which the enol is formed.) In the absence of an added acid,

110

the formation of propene-2-ol from propanone – to replace that removed in the reaction with iodine – is extremely slow, being dependent for catalysis on H^+ ions arising from the dissociation of H_2O molecules:

$$H_2O \rightleftharpoons H^+ + OH^-$$

At neutral pH values, the concentration of H^+ is, of course, tiny (1.00×10^{-7} mol dm^{-3} at 25 °C). Under such conditions, with the concentration of propanone taken to be 1.00 mol dm^{-3}, the rate of enol formation would be impossibly slow,

$$\text{rate} = k_E[H^+][CH_3COCH_3]$$

$$= (2.79 \times 10^{-5}\,\text{dm}^3\,\text{mol}^{-1}\,\text{s}^{-1})\,(1.00 \times 10^{-7}\,\text{mol dm}^{-3})\,(1.00\,\text{mol dm}^{-3})$$

$$= 2.8 \times 10^{-12}\,\text{mol dm}^{-3}\,\text{s}^{-1}$$

which means there would be a long 'delay' before there would be enough enol to react with the remaining iodine (present in excess). If, however, the iodination of propanone is carried out in the presence of an *added* acid, the rate of enol formation from propanone is much greater. At a H^+ concentration of 0.250 mol dm^{-3}, for example, the reaction is 2.5-million times faster:

$$\text{rate} =$$

$$(2.79 \times 10^{-5}\,\text{dm}^3\,\text{mol}^{-1}\,\text{s}^{-1})\,(0.250\,\text{mol dm}^{-3})\,(1.00\,\text{mol dm}^{-3})$$

$$= 7.0 \times 10^{-6}\,\text{mol dm}^{-3}\,\text{s}^{-1}$$

It is now easy to see why the acid-catalysed formation of propanone to propene-2-ol is the rate-determining step in the *overall* reaction between iodine and propanone: when measuring the rate of propanone iodination, what you are really measuring is the rate of propanone conversion to propene-2-ol. In other words, the rate constant for reaction between iodine and propanone is same as that for the reaction between propanone and the H^+ ion (k_E, which equals 2.79×10^{-5} dm^3 mol^{-1} s^{-1}) and is miniscule compared with that for the reaction of iodine with propene-2-ol (4.0×10^9 dm^3 mol^{-1} s^{-1}).

Although increasing the concentration of H^+ increases the *rate* of propanone conversion to its enol form, **the catalyst cannot increase the concentration of the enol to a value above that prescribed by the equilibrium constant** ($K_c = 4.69 \times 10^{-9}$). It will be instructive here to examine this important statement in some detail. For any reversible reaction, a catalyst increases the rate of both the forward and reverse reactions by the *same* factor. The overall equation for the acid-catalysed isomerisation of propanone between its ketone and enol forms is, as we have seen above,

$$CH_3COCH_3 \; + \; H^+ \; \rightleftharpoons \; CH_3C(OH)CH_2 \; + \; H^+$$

and may be written as the separate 'forward' and 'reverse' reactions, each with its own rate equation,

$$CH_3COCH_3 \; + \; H^+ \; \rightarrow \; CH_3C(OH)CH_2 \; + \; H^+$$

$$\text{rate}_{(forward)} \; = \; k_E \, [H^+][CH_3COCH_3]$$

and

$$CH_3C(OH)CH_2 \; + \; H^+ \; \rightarrow \; CH_3COCH_3 \; + \; H^+$$

$$\text{rate}_{(reverse)} \; = \; k_K \, [H^+][CH_3C(OH)CH_2]$$

where k_E, as we have seen, is the rate constant for the formation of the enol and k_K is the rate constant for the reverse reaction, *i.e.* for the formation of propanone in its ketone form from propene-2-ol. At a value of 5.95×10^3 dm^3 mol^{-1} s^{-1}, k_K is considerably greater than k_E (2.79×10^{-5} dm^3 mol^{-1} s^{-1}).

It was shown above that increasing the concentration of H^+ in a 1.00 mol dm^{-3} solution of propanone, from 1.00×10^{-7} to 0.250 mol dm^{-3}, increases the rate of its conversion to propene-2-ol by a factor of 2.5-million (from 2.8×10^{-12} to 7.0×10^{-6} mol dm^{-3} s^{-1}). Let's now look at how the same increase in the concentration of H^+ affects the rate of the reverse reaction. Since K_c equals 4.69×10^{-9}, the concentration of propene-2-ol in our 1.00 mol dm^{-3} solution of propanone will be 4.69×10^{-9} mol dm^{-3}. In the absence of an added acid, where $[H^+]$ is taken to be 1.00×10^{-7} mol dm^{-3}, the rate of propanone formation from propene-2-ol is, then:

$$\text{rate} = k_E [H^+][CH_3C(OH)CH_2]$$

$$= (5.95 \times 10^3 \, dm^3 \, mol^{-1} \, s^{-1})(1.00 \times 10^{-7} \, mol \, dm^{-3})$$

$$(4.69 \times 10^{-9} \, mol \, dm^{-3})$$

$$= 2.8 \times 10^{-12} \, mol \, dm^{-3} \, s^{-1}$$

Notice how this rate is identical to that of the forward reaction when the concentration of the H^+ ion is also 1.00×10^{-7} mol dm^{-3}. This, of course, is to be expected because the concentrations of propanone and propene-2-ol used in both calculations are the equilibrium concentrations and, by definition, when a reaction is at equilibrium its forward and reverse reactions are proceeding at identical rates.

Although increasing the concentration of the H^+ ion to 0.25 mol dm^{-3} increases the rate of the forward reaction, it increases the rate of the reverse reaction *by exactly the same factor* (2.5-million), bringing it up to the same value as that of the forward reaction:

$$\text{rate} = (5.95 \times 10^3 \, dm^3 \, mol^{-1} \, s^{-1}) (0.250 \, mol \, dm^{-3})$$

$$(4.69 \times 10^{-9} \, mol \, dm^{-3})$$

$$= 7.0 \times 10^{-6} \, mol \, dm^{-3} \, s^{-1}$$

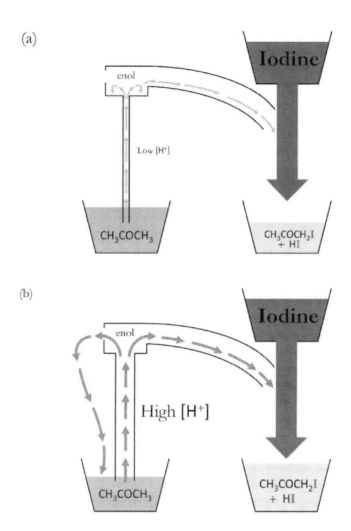

(a)

(b)

Figure 3.2 The overall rate at which iodine reacts with propanone is limited by the availability of the enol form of propanone (propene-2-ol), formed in the reaction between propanone and the H^+ ion (a). Increasing the concentration of H^+ increases the rate at which the enol – that has reacted with iodine – is replaced by the isomerisation of more propanone. For any given amount of propanone, there is a limit to the amount of the enol that can be formed (the ratio [propene-2-ol]/[propanone] cannot exceed 4.69 $\times$ 10^{-9} at 25 °C). This is because H^+ ions also catalyse the isomerisation of propene-2-ol back to propanone, as indicated by the enol 'overspill' in (**b**). See text for further details.

It is seen, therefore, that while increasing the concentration of H^+ increases the rate at which propene-2-ol is generated from propanone, this does not increase the concentration of propene-2-ol attainable because the additional H^+ ions also increase the rate of propene-2-ol conversion back to the ketone. The reason increasing the concentration of H^+ increases the rate of the overall reaction between iodine and propanone is because the ion increases the rate at which the enol – removed in its rapid reaction with iodine – is replaced through the isomerisation of more propanone. This is shown schematically in **Figure 3.2**.

Taking the above reasoning slightly further shows how the equilibrium constant of a reversible reaction is a function of the rate constants of the forward and reverse reactions. At equilibrium, the rates of the forward and reverse reactions are, by definition, equal:

$$rate_{(forward)} \quad = \quad rate_{(reverse)}$$

Substitution of the respective rate equations gives,

$$k_E[H^+][CH_3COCH_3] \quad = \quad k_K[H^+][CH_3C(OH)CH_2]$$

which, after cancelling out the $[H^+]$ term on both sides, we can rearrange to:

$$\frac{k_E}{k_K} \quad = \quad \frac{[CH_3C(OH)CH_2]}{[CH_3COCH_3]}$$

The right-hand side of this expression corresponds to the equilibrium constant, the reported value of which was obtained from the experimentally determined rate constants, k_E ($2.79 \times 10^{-5}\,dm^3\,mol^{-1}\,s^{-1}$) and k_K ($5.95 \times 10^3\,dm^3\,mol^{-1}\,s^{-1}$):[aaa]

$$\frac{k_E}{k_K} \quad = \quad \frac{2.79 \times 10^{-5}\,mol\,dm^{-3}}{5.95 \times 10^3\,mol\,dm^{-3}}$$

$$= \quad 4.69 \times 10^{-9}$$

[aaa] See *Further reading* (article by Chiang, Kresge and Schepp, 1989).

The presence of a catalyst does not change the fact that the concentration of propene-2-ol in a 1.0 mol dm^{-3} solution of propanone at equilibrium is miniscule (only 4.69×10^{-9} mol dm^{-3}): catalysts do not change the values of equilibrium constants. By increasing the rate of both the forward and reverse reactions, the presence of a catalyst results in a reversible reaction reaching equilibrium sooner.

3.3   The Harcourt-Essen reaction

The publication in the 1860s of a series of seminal papers by A. Vernon Harcourt and William Essen can be said to mark the birth of the field of reaction kinetics. A major focus of their studies was the oxidation of iodide ions by hydrogen peroxide in the presence of hydrogen ions:

$$H_2O_2 \quad + \quad 2\,I^- \quad + \quad 2\,H^+ \quad \rightarrow \quad 2\,H_2O \quad + \quad I_2$$

The Harcourt-Essen reaction is a popular practical in A-level chemistry courses, not least because it lends itself to investigation using the so-called 'clock method'. Details of the methods used to measure reaction rates in school laboratories are to be covered in the next chapter. The focus of our attention in the present chapter are the links between rate equations and reaction mechanisms. In the majority of textbooks the Harcourt-Essen reaction is reported to display second-order kinetics, obeying the rate equation:[bbb]

$$\text{rate} \quad = \quad k_2\,[H_2O_2][I^-]$$

This reflects the rate-determining step, in which an oxygen atom is transferred from the peroxide to the iodide ion, giving the iodate(I) ion.

$$H_2O_2 \quad + \quad I^- \quad \rightarrow \quad H_2O \quad + \quad IO^- \qquad \text{(slow)}$$

[bbb] To indicate that the rate constant is for a second-order reaction, it is being written here as k_2, with a subscripted 2.

The IO^- ion is the conjugate base of iodic(I) acid.[ccc] As IOH is a very weak acid,[ddd] the IO^- ion then accepts a proton:

$$IO^- + H^+ \rightleftharpoons HOI \qquad \text{(fast)}$$

In the final reaction step, which is also fast, HOI oxidises the second iodide ion:

$$HOI + I^- + H^+ \rightarrow I_2 + H_2O \qquad \text{(fast)}$$

Adding together the three reaction steps, and cancelling out HOI and the IO^- ion, gives the overall equation for the reaction:

$$H_2O_2 + 2\,I^- + 2\,H^+ \rightarrow 2\,H_2O + I_2$$

The Harcourt-Essen reaction is found to obey the second-order rate equation shown above only if the concentration of H^+ ions is low; otherwise, the reaction is observed to follow more complex kinetics, in which the rate *is* affected by the concentration of the H^+ ion. This is because the reaction proceeds by two different mechanisms *at the same time* – each with a different rate-determining step.

The second of the two mechanisms, which is first order with respect to the H^+ ion, has the following rate-determining step,

$$H_2O_2 + I^- + H^+ \rightarrow H_2O + HOI \qquad \text{(slow)}$$

being followed by the same fast, final step as the second-order mechanism:

$$HOI + I^- + H^+ \rightarrow I_2 + H_2O \qquad \text{(fast)}$$

[ccc] Although perhaps not familiar with these species, you are likely to have encountered their chlorine equivalents, the chlorate(I) ion (ClO^-) and chloric(I) acid (HOCl), in your first-year A-level studies. Most textbooks give the chemical formula of chloric(I) acid as HClO, but this is not strictly correct: studies in the gas phase have shown that HClO (in which the hydrogen atom is bonded to chlorine, rather than oxygen) is an unstable isomer of HOCl. HClO is around $280\,kJ\,mol^{-1}$ higher in energy than HClO – they are different chemical species (Turner, G., 1986, *Inorg. Chim. Acta* **111**, 157 – 161). For this reason, the formula of iodic(I) acid used here is HOI.

[ddd] The acid dissociation constant, K_a, for HOI is $1.3 \times 10^{-10}\,mol\,dm^{-3}$, making it a much weaker acid than the more familiar weak acids, the carboxylic acids, covered at A-level.

It obeys the third-order rate equation:

$$\text{rate} \;=\; k_3\,[H_2O_2][I^-][H^+]$$

As it is the mechanism of the Harcourt-Essen reaction that is given in the majority of textbooks, you might expect the former mechanism, with the second-order rate equation, to be the more important, but its rate constant (k_2), 0.0115 dm^3 mol^{-1} s^{-1}, is considerably smaller than that for the third-order reaction ($k_3 = 0.175$ dm^6 mol^{-2} s^{-1}). The overall rate of the Harcourt-Essen reaction – observed under any given set of reactant concentrations – is simply the sum of the rates of the reactions proceeding by the two, independent mechanisms,

$$\text{overall rate} \;=\; k_2\,[H_2O_2][I^-] \;+\; k_3\,[H_2O_2][I^-][H^+]$$

We can now illustrate why the reaction appears to be zero order with respect to H$^+$ when the concentration of the ion is low. Firstly, we will calculate the overall rate (usually defined as the rate of iodine formation) with all three reactants at an initial concentration of 0.525 mol dm^{-3}. As these are the concentrations of the reactants only at the start of the reaction, the calculated value will be the initial rate:

$$\text{initial rate} \;=\; (0.0115 \text{ dm}^3 \text{ mol}^{-1} \text{ s}^{-1})(0.525 \text{ mol dm}^{-3})\,(0.525 \text{ mol dm}^{-3})$$

$$+\;\; (0.175 \text{ dm}^6 \text{ mol}^{-2} \text{ s}^{-1})(0.525 \text{ mol dm}^{-3})$$

$$(0.525 \text{ mol dm}^{-3})(0.525 \text{ mol dm}^{-3})$$

$$=\;\; 3.17 \times 10^{-3} \text{ mol dm}^{-3} \text{ s}^{-1} \;+\; 2.53 \times 10^{-2} \text{ mol dm}^{-3} \text{ s}^{-1}$$

$$=\;\; 2.85 \times 10^{-2} \text{ mol dm}^{-3} \text{ s}^{-1}$$

At these reactant concentrations, it is seen that the third-order mechanism, *involving the H$^+$ ion in its rate-determining step*, makes a far greater contribution to the overall rate than the second-order mechanism. Under such conditions, the observed rate would not be independent of the concentration of the H^+ ion – in other words, we would not observe zero-order behaviour in H^+. Now consider the situation in which the concentration of the H^+ ion is lowered to 5.00×10^{-4} mol dm^{-3}:

$$\text{initial rate} = (0.0115 \text{ dm}^3 \text{ mol}^{-1} \text{ s}^{-1})(0.525 \text{ mol dm}^{-3})(0.525 \text{ mol dm}^{-3})$$
$$+ \quad (0.175 \text{ dm}^6 \text{ mol}^{-2} \text{ s}^{-1})(0.525 \text{ mol dm}^{-3})$$
$$(0.525 \text{ mol dm}^{-3})(5.00 \times 10^{-4} \text{ mol dm}^{-3})$$

$$= \quad 3.17 \times 10^{-3} \text{ mol dm}^{-3} \text{ s}^{-1} \quad + \quad 2.41 \times 10^{-5} \text{ mol dm}^{-3} \text{ s}^{-1}$$
$$= \quad 3.19 \times 10^{-3} \text{ mol dm}^{-3} \text{ s}^{-1}$$

The reaction proceeding *via* the third-order mechanism, involving the H^+ ion, is seen to make only a tiny (less than 1 %) contribution to the overall rate. Under these conditions, we would expect changes in the concentration of H^+ to have essentially no observable effect on the rate. Thus, doubling $[H^+]$ to 1.00×10^{-3} mol dm^{-3} would cause the rate to increase to only 3.22×10^{-3} mol dm^{-3} s^{-1}, which is well within the experimental error achieved in school laboratories!

In a variant of the Harcourt-Essen reaction that is used widely in teaching, the peroxodisulfate(VI) ion, $S_2O_8^{2-}$, is used in place of hydrogen peroxide as the oxidising agent:

$$S_2O_8^{2-} \quad + \quad 2 I^- \quad \rightarrow \quad 2 SO_4^{2-} \quad + \quad I_2$$

This reaction is a Core Practical in the current Edexcel specification[eee] and is included in the AQA course to illustrate the catalysis of a reaction by a transition-metal ion (Fe^{2+}), which we shall consider below.[fff] It is very important, therefore, that you are familiar with the peroxodisulfate(VI) ion, which has the structure:

$$-2 \ \ddot{\text{O}} = \overset{\overset{\displaystyle -2}{\overset{\displaystyle :\ddot{\text{O}}:}{\|}}}{\underset{\underset{\displaystyle -2}{\underset{\displaystyle :\text{O}:}{|}}}{\text{S}}} - \overset{-1}{\ddot{\text{O}}} - \overset{-1}{\ddot{\text{O}}} - \overset{\overset{\displaystyle -2}{\overset{\displaystyle :\ddot{\text{O}}:}{\|}}}{\underset{\underset{\displaystyle -2}{\underset{\displaystyle :\text{O}:}{|}}}{\text{S}}} = \ddot{\text{O}}: \ -2$$

The numbers are the oxidation states (oxidation numbers) of the eight oxygen atoms. You should know from your studies of oxidation states that, when bonded to other atoms, oxygens are in the −2 state, except in peroxides (−O−O−), such as H_2O_2, where each oxygen is −1. The $S_2O_8{}^{2-}$ ion is also a peroxide, in which the two bridging oxygen atoms (shown in red) are in the −1 state; the remaining 6 oxygens are in the −2 state.

When the $S_2O_8{}^{2-}$ ion accepts a pair of electrons to form two sulfate ions, its two peroxidic oxygens are reduced from −1 to −2 (all four oxygens in $SO_4{}^{2-}$ are in the −2 state). The two sulfur atoms are in the +6 oxidation state throughout – hence, the names peroxodisulfate(VI) and sulfate(VI).

The reduction of peroxodisulfate(VI) by iodidie ions is a second-order reaction, with the following rate equation:

$$\text{rate} \quad = \quad k \ [S_2O_8{}^{2-}][I^-]$$

[eee] Edexcel Specification, Appendix 5d. Core Practical 14: Finding the activation energy of a reaction.
[fff] AQA Specification, Section 3.2.5.6 Catalysts.

The rate-determining step involves an encounter between the two ions, which is believed to proceed as follows:[ggg]

$$:\ddot{O}=\overset{\overset{\textstyle :\ddot{O}:}{\|}}{\underset{\underset{\textstyle :\ddot{O}:}{|}}{S}}-\ddot{O}-\ddot{O}-\overset{\overset{\textstyle :\ddot{O}:}{\|}}{\underset{\underset{\textstyle :\ddot{O}:}{|}}{S}}=\ddot{O}: \quad\quad [\ :\ddot{I}:\ ]^{-}$$

$$\downarrow$$

$$:\ddot{O}=\overset{\overset{\textstyle :\ddot{O}:}{\|}}{\underset{\underset{\textstyle :\ddot{O}:}{|}}{S}}-\ddot{O}:^{-} \quad + \quad {}^{-}:\ddot{O}-\overset{\overset{\textstyle :\ddot{O}:}{\|}}{\underset{\underset{\textstyle :\ddot{O}:}{|}}{S}}=\ddot{O}: \quad + \quad [\ :\ddot{I}:\ ]^{+}$$

Notice how the oxygen atom on the right of the peroxide bridge takes the two electrons from the O–O bond. At the same time, the oxygen on the left accepts two electrons from the I^- ion, which becomes an I^+ ion. These electron movements are often shown to occur within a complex, formed between the two ions:

$$S_2O_8{}^{2-} \ + \ I^- \ \rightarrow \ S_2O_8I^{3-} \ \rightarrow \ 2\ SO_4{}^{2-} \ + \ I^+$$

This slow, rate-determining step is then followed by the very fast reaction of I^+ with the *second* I^- ion, forming an iodine molecule:

$$I^+ \ + \ I^- \ \rightarrow \ I_2$$

For peroxodisulfate(VI) and iodide ions to form a transition state in the rate-determining step, they must collide with sufficient energy to overcome the electrostatic repulsion between their negative charges.

[ggg] Other mechanisms have been proposed, including one involving iodine radicals. For a review of this reaction, see D. A. House (1962) Kinetics and mechanism of oxidations by peroxydisulfate. *Chem. Rev.* **62**, 197– 203.

As we saw in the previous chapter, the energy required to form the transition state corresponds to the activation energy of a reaction, which for this reaction is approximately 50 kJ mol^{-1}.[hhh]

The oxidation of iodide ions by peroxodisulfate(VI) ions is catalysed by Fe^{2+} or Fe^{3+} ions. If Fe^{2+} ions are added, they are rapidly oxidised by the peroxide:

$$S_2O_8{}^{2-} + 2\,Fe^{2+} \rightarrow 2\,SO_4{}^{2-} + 2\,Fe^{3+}$$

In this reaction, each Fe^{2+} ion releases one electron. The two electrons are then fed into the O–O bond, forming two sulfate ions. The reaction can be understood more easily by *imagining* that the O–O bond – shown below as the shaded electron pair – first breaks by homolytic fission to form two radicals, each of which then receives an electron to become a sulfate ion:

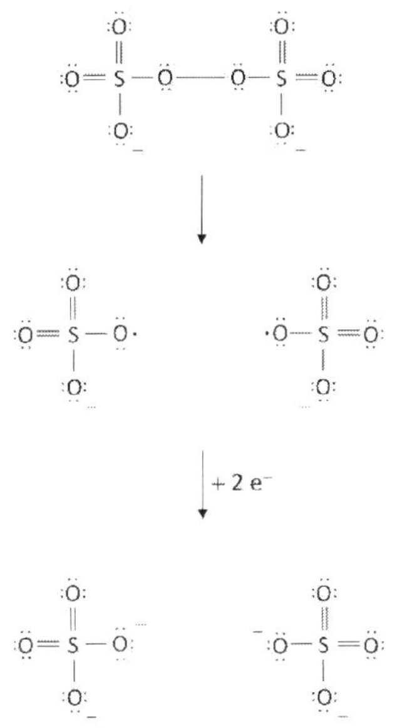

The rate constant for this reaction is 5.0×10^3 dm^3 mol^{-1} s^{-1} at 25 °C. Notice, from the units, that this is a second-order rate constant, reflecting a two-step mechanism: the slower, rate-determining step

---

[hhh] The measurement of the activation energy of this reaction is described in Chapter 4.

.

involves a collision between peroxodisulfate(VI) and just one Fe^{2+} ion, in which only one of the oxygen atoms from the O–O bond receives an electron to become a sulfate ion. The other 'half' of the peroxide forms a radical, which then – in a much faster reaction step – oxidises the second Fe^{2+} ion:

$$S_2O_8{}^{2-} \ + \ Fe^{2+} \ \rightarrow \ SO_4{}^{2-} \ + \ SO_4{}^{\cdot-} \ + \ Fe^{3+} \quad \text{(slow)}$$

$$SO_4{}^{\cdot-} \ + \ Fe^{2+} \ \rightarrow \ SO_4{}^{2-} \ + \ Fe^{3+} \quad\quad\quad \text{(fast)}$$

Then, in a separate reaction, Fe^{3+} ions oxidise iodide ions to iodine, thereby regenerating Fe^{2+}:

$$2\ Fe^{3+} \ + \ 2\ I^- \ \rightarrow \ I_2 + \ 2\ Fe^{2+}$$

Although the equation shows two Fe^{3+} ions reacting with two iodide ions, this reaction is third order, with the rate equation:

$$\text{rate} \quad = \quad k\ [Fe^{3+}][I^-]^2$$

This relatively slow reaction ($k = 16$ dm^6 mol^{-2} s^{-1} at 25 °C) is believed to reflect the formation of the radical species $I_2{}^{\cdot-}$, which is nothing more than an iodine molecule with an added electron,

$$Fe^{3+} \ + \ I^- \ \rightleftharpoons \ Fe^{2+} \ + \ I\cdot$$

$$I\cdot \ + \ I^- \ \rightleftharpoons \ I_2{}^{\cdot-}$$

where $I\cdot$ is an iodine atom (the 'radical dot' representing an unpaired electron). These two reaction steps are believed to occur within a complex formed between Fe^{3+} and I^-, namely $[FeI]^{2+}$. The iodide ion transfers an electron to Fe^{3+} *within* the complex, forming Fe^{2+} and $I\cdot$, which adds to the second iodide ion (without the release of $I\cdot$ as a 'free' atom). Although this process is perhaps easier to understand when written as the two equations shown above, a more accurate description would be,

$$Fe^{3+} \ + \ I^- \ \rightleftharpoons \ [FeI]^{2+}$$

$$[FeI]^{2+} \ + \ I^- \ \rightleftharpoons \ Fe^{2+} \ + \ I_2{}^{\cdot-}$$

which is, indeed, how the reaction is given in the research literature.[iii] The $I_2^{•-}$ species then rapidly transfers its extra electron to a second Fe^{3+} ion,

$$I_2^{•-} \quad + \quad Fe^{3+} \quad \rightarrow \quad I_2 \quad + \quad Fe^{2+}$$

thereby regenerating Fe^{2+} and completing the overall catalytic cycle of iodide oxidation by peroxodisulfate(VI). The rate-determining step in the reduction of Fe^{3+} by I^- is the formation of $I_2^{•-}$. This requires the interaction of two iodide ions and one Fe^{3+} ion (in the two reversible reactions shown above), thereby accounting for the third-order rate equation.

It is stated in the AQA specification (2015 onwards) that students are expected to be able to explain, with the aid of equations, how Fe^{2+} ions catalyse the reaction between I^- and $S_2O_8^{2-}$. The current Edexcel specification states only that a knowledge of the process is required. In none of the textbooks endorsed by these boards, however, are the reactions described in the level of detail given here. The AQA textbook from the Oxford University Press, for example, simply shows two steps in the catalytic cycle,

$$S_2O_8^{2-} \quad + \quad 2\,Fe^{2+} \quad \rightarrow \quad 2\,SO_4^{2-} \quad + \quad 2\,Fe^{3+}$$

$$2\,Fe^{3+} \quad + \quad 2\,I^- \quad \rightarrow \quad I_2 + \quad 2\,Fe^{2+}$$

explaining that both involve the collision of oppositely-charged ions, thereby circumventing the requirement for collisions between negatively-charged ions (between which there is electrostatic repulsion) in the uncatalysed reaction.

The current OCR specification mentions the catalytic properties of the transition elements, but states that no detail of catalytic processes is required. For such reasons, I often ask myself whether in my teaching and tutoring I am not overloading students with too much detail. Every now and then, however, an exam question comes along that tells me that this is not the case. Question 17 in the 2018 A-level Paper 1 from the OCR board (Chemistry A) gives initial rates data for the reaction between Fe^{3+} and I^- ions, from which candidates are asked to derive the rate equation and rate constant. No worries there, but they are then

---

[iii] G. S. Laurence and K. J. Ellis (1972) Oxidation of iodide ion by Fe(III) in aqueous solution. *Journal of the Chemical Society, Dalton Transactions*, 2229 – 2233.

asked to propose a mechanism based on the rate equation. The various mechanisms given in the mark scheme are essentially different versions of the mechanism given here, involving the $I_2^{\cdot-}$ species (albeit with the 'radical dot' missing). I was first shown this question by a student I was then tutoring. It had appeared in his mock examination, which we were reviewing together. I remember thinking, how on earth is an A-level student expected to be aware that a species such as $I_2^{\cdot-}$ can be formed? The examination board would probably say that students are expected to 'apply the concepts' covered in the course. This is true, but it takes a lot of confidence on the part of a student to propose the involvement of a chemical species that looks as alien as the iodine radical anion (its formal name) – a species I recognised as a powerful reducing agent from my postdoctoral work on free radicals and radiation chemistry, where I was familiar with the analogous chlorine radical anion.

The fact of the matter is that the textbooks endorsed by the examination boards can give neither the depth nor detail needed to deal with the more difficult examination questions, which are intended to filter out the strongest candidates, whose understanding of the key concepts is sufficiently secure for them to be able to apply them to unfamiliar reactions – either that or their teachers have expanded on the material (which, knowing how pushed teachers are for time, is what I am doing for you here).

Before moving on to the consideration – justifiably, in some detail – of one or two other reactions that may appear in your examinations, it should be pointed out that the reaction between peroxodisulfate(VI) and iodide ions can also be described as being catalysed by Fe^{3+} ions. Although Fe^{2+} has been shown to be the catalyst in the description given above, the same effect can be achieved by adding Fe^{3+} ions, in which case iodide ions first reduce the metal ion, followed by the reaction of Fe^{2+} with $S_2O_8^{2-}$, thereby regenerating the Fe^{3+} catalyst. In either case, the reduction of Fe^{3+} by the iodide ion is the slower step, as reflected in relative sizes of the respective rate constants.

3.4   Isomerisation of cyclopropane

In the previous chapter, the procedure for obtaining the activation energy (and Arrhenius constant) of a reaction using the Arrhenius equation was explained using the decomposition of ethanal to methane and carbon monoxide as an example:

$$CH_3CHO \rightarrow CH_4 + CO$$

Although the balanced equation for this reaction shows a single molecule of ethanal decomposing into one molecule of each of the products, the rate equation indicates that the reaction is second order with respect to the reactant:

$$\text{rate} = k [CH_3CHO]^2$$

Compare this reaction with the isomerisation of cyclopropane to propene:

Again, the balanced equation shows the reaction of a single molecule of reactant, but in this case the reaction is first order, with the following rate equation:

$$\text{rate} = k [cyclo\text{-}C_3H_6]$$

Think carefully about what these rate equations are telling us about the mechanisms of the two reactions: whereas the rate-determining step of the former involves a collision between two molecules of ethanal, that of the latter reaction involves only a single molecule of cyclopropane.

As we saw in the previous chapter, the activation energy of a reaction reflects the energy that must be gained by the reactants to form the transition state (sitting at the point of highest energy in the reaction profile). According to collision theory, reactants acquire this energy through their collisions with each other. Thus, we have seen how the exponential term in the Arrhenius equation enables us to calculate the fraction of collisions in a population of reactant molecules – at a given temperature – that occur with an energy of at least the activation energy. This is easy enough for us to envisage in the second-order decomposition of ethanal – only collisions between pairs of ethanal molecules that occur with an energy of at least 180 kJ mol^{-1} can result in a reaction (see Chapter 2) – but poses a dilemma when dealing with first-order reactions: if the rate-determining step in the isomerisation of cyclopropane involved a pair of colliding molecules, the reaction would be second order. How, then, does a cyclopropane molecule acquire the extra energy needed to break itself open, forming propene? The fact that the rate constant of the reaction increases with increasing temperature confirms that the isomerisation of cyclopropane molecules to propene does indeed involve their gaining energy.

The explanation to this apparent dilemma requires us to remember that rate equations and rate constants are based on experimental observations – what reactions 'allow us to see' – and often reflect more fundamental, underlying processes. In the 1920s, Fredrick Lindemann proposed a theory to explain how the reactant in a reaction that displays first-order kinetics acquires the energy needed to react. According to this model, the isomerisation of cyclopropane to propene involves the popping open of a single, 'energised' molecule of cyclopropane, which we can represent by:

$$(cyclo\text{-}C_3H_6)^* \quad \rightarrow \quad \text{propene}$$

The energised molecule, $(cyclo\text{-}C_3H_6)^*$, arises during the random collisions between cyclopropane molecules, therefore *its* generation is not a first-order process. If we assume that the formation of $(cyclo\text{-}C_3H_6)^*$ is much faster than its subsequent isomerisation to propene, then, according to the Lindemann theory, the overall reaction will appear to be first order. The details of this reasoning are beyond the scope of A-level chemistry, but they are given in **Appendix II** for those wishing to extend their knowledge of the subject.

3.5   Decomposition of hydrogen peroxide

The rate of hydrogen peroxide decomposition is often investigated in school practicals, both at GCSE and A-level. The reaction produces oxygen gas, which is simple to measure using a gas syringe or an inverted measuring cylinder filled with water.

$$2\,H_2O_2 \quad \rightarrow \quad 2\,H_2O \quad + \quad O_2$$

As well as being studied in rates-of-reaction modules, the decomposition of hydrogen peroxide is of interest in A-level courses because it is an example of a disproportionation reaction[iii].

The decomposition of hydrogen peroxide is, in fact, an extremely slow reaction. Aqueous solutions of the reagent are often stored for several years in laboratory refrigerators, where the H_2O_2 undergoes only minimal decomposition. In your GCSE studies, you may have investigated the catalysis of hydrogen peroxide decomposition by manganese(IV) oxide or by pieces of liver or celery, which contain enzymes that catalyse the reaction. Even when a catalyst has not been added, it is almost impossible to prevent the catalysis of H_2O_2 decomposition by metal ions – present at trace levels – or by the surface of the vessel in which the solution is contained. Such reactions are responsible for the slow breakdown that occurs in stored solutions of the reagent.

In the *absence* of a catalyst, the decomposition begins with the homolytic fission of the O–O single bond in hydrogen peroxide, forming two hydroxyl radicals:

$$H—\ddot{O}——\ddot{O}—H$$

$$\downarrow$$

$$H—\ddot{O}\cdot \qquad \cdot\ddot{O}—H$$

Under normal laboratory conditions, the rate of homolytic fission of the O–O bond is extremely low, resulting in only a barely detectable rate of H_2O_2 decomposition[kkk]. At elevated temperatures and pressures, however, the reaction is fast.

The $\cdot$OH radical is a very powerful oxidising agent. Its generation during the decomposition of hydrogen peroxide at high temperatures and pressures has been investigated as a means of degrading hazardous chemicals, present in industrial wastes, in a process known as supercritical water oxidation. When water in a closed container is heated,

---

[iii] The O atoms in H_2O_2 are in the −1 oxidation state (their oxidation number is −1), being both reduced to −2 in H_2O and oxidised to +1 in O_2.

[kkk] Hydrogen peroxide does undergo decomposition when exposed to ultraviolet radiation, including that at the wavelengths present in sunlight. The energy carried by the radiation is sufficient to cause homolytic fission of the O–O bond (a process known as photolysis).

the vapour formed as the liquid evaporates cannot escape, resulting in an increase in pressure. This pressure, pushing back down on the remaining liquid, makes it increasingly difficult for more water molecules to escape into the gas phase, allowing the temperature of the liquid to increase to far beyond the temperature at which it would otherwise boil (100 °C at atmospheric pressure). Eventually, when the temperature reaches 374 °C, enough water molecules have entered the gas phase for its pressure to equal the pressure caused by water molecules evaporating from the surface of the liquid. At this point, the boundary between the two phases disappears and a single, homogenous phase – known as a supercritical fluid – is formed. The temperature and pressure at which this occurs, 374 °C and 2.21×10^7 Pa, are together called the **critical point of water** (2.21×10^7 Pa is over 200-times atmospheric pressure).

In the form of a supercritical fluid, water assumes properties that make it a suitable medium for the degradation of hazardous chemicals by oxidising agents, including hydrogen peroxide. In particular, the polarity of water decreases when it reaches its critical point, resulting in the formation of a homogenous phase containing organic compounds, including hydrocarbons and polychlorinated biphenyls (BCPs), that are immiscible with water in its liquid state.

Figure 3.3 shows the rapid decomposition of hydrogen peroxide dissolved in water above its critical point. To determine how changing the concentration of H_2O_2 affects the reaction rate, and hence the order of the reaction, a series of tangents may be drawn to the curve at selected concentrations of the peroxide. The tangent shown in the figure was drawn to the curve when the H_2O_2 concentration was 0.0200 mol dm^{-3}, to which it had fallen after 0.783 seconds. The rate is equal to the gradient of the tangent:

$$\text{rate} \;=\; \frac{0.0383 \ \text{mol dm}^{-3}}{1.64 \ \text{s}}$$

$$=\; 0.0234 \ \text{mol dm}^{-3} \ \text{s}^{-1}$$

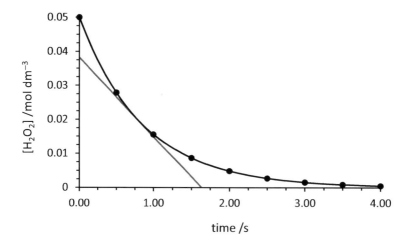

Figure 3.3   Time course of hydrogen peroxide decomposition at 400 °C and 2.45 × 10⁷ Pa. The data points were calculated using kinetic parameters for the reaction taken from a study by Eric Croiset, Steven F. Rice and Russell G. Hanush (*AIChE Journal*, 1997, **43**, 2343 – 2352). The rate of the reaction at a H_2O_2 concentration of 0.0200 mol dm⁻¹ was calculated from the tangent drawn to the curve at this point. See text for further details.

time/s	$[H_2O_2]$/ mol dm^{-3}	rate/ mol dm^{-3} s^{-1}
0.191	0.0400	0.0468
0.437	0.0300	0.0351
0.783	0.0200	0.0234
1.376	0.0100	0.0117

Table 3.3   Rates of hydrogen peroxide decomposition obtained from tangents drawn to the curve shown in **Figure 3.3** after the concentration of H_2O_2 had fallen to various levels. For further details, see the text and the legend to **Figure 3.3**.

The rates obtained from gradients drawn to the curve after the concentration of H_2O_2 had fallen to a range of different values are given in **Table 3.3**. The effect of changing the H_2O_2 concentration on the rate can be seen by comparing any two pairs of values. We see, for example, that increasing the concentration from 0.0200 to 0.0400 mol dm^{-3} doubles the rate. Similarly, increasing the H_2O_2 concentration 3-fold (from 0.0100 to 0.0300 mol dm^{-3}) causes the rate to increase 3-fold (from 0.0117 to 0.0351 mol dm^{-3} s^{-1}). This indicates that the reaction is first order with respect to H_2O_2.

In examination questions, you are often asked to determine the order of a reaction (with respect to the concentration of a particular reactant) from data presented in a table. Where there are two or more reactants involved, it is of course necessary to compare pairs of values where the concentration of only one reactant has been changed, allowing you to determine its effect on the rate in isolation from any effects caused by changing the concentrations of the other reactants. An example of this was given in Chapter 1, where the rates of the reaction between BrO_3^-, Br^- and H^+ ions were compared at various concentrations of each reactant.

Having concluded that the decomposition of hydrogen peroxide is a first-order reaction, we can write the rate equation:

$$\text{rate} = k\,[H_2O_2]$$

We could find the value of the rate constant simply by substituting into the rate equation any pair of values from **Table 3.3**:

$$k = \frac{\text{rate}}{[H_2O_2]}$$

A more accurate method, however, is to obtain k from a plot of rate against H_2O_2 concentration. This is because the rate equation is of the form $y = mx + c$, with c equal to zero; in other words, it is the equation of a straight line. A plot of rate (on the y-axis) against hydrogen peroxide concentration (x-axis) will give a straight line, passing through the origin (c is zero) and with a gradient (m) equal to k. This method is more accurate than the former because it averages out the slightly different

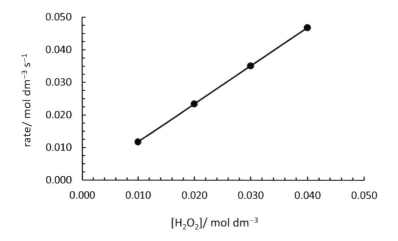

Figure 3.4 Rate versus concentration plot of the data given in **Table 3.3**. The first-order rate constant for the decomposition of hydrogen peroxide ($1.17 \ s^{-1}$ at 400 °C and 2.45×10^7 Pa) was obtained by calculating the gradient of the straight line. See text for details.

values of k that might be obtained using individual pairs of values given in the table.[III] Such a plot is shown in **Figure 3.4**. The fact that the graph is a straight line confirms that the reaction is first order. Its gradient, which equals k, is $1.17 \ s^{-1}$. Notice how the units of the first-order rate constant are obtained by cancelling out the units of rate in the numerator with those of concentration in the denominator. This is why it is always good practice to include units within your calculations, treating them just as you would numbers. You should not be expending your mental energies trying to memorise the units of first-, second- and third-order rate constants!

Having derived the rate equation, it is now necessary to propose a reaction mechanism with which it is consistent. Although the equation for the overall reaction shows the reaction between two molecules of H_2O_2,

$$2 \ H_2O_2 \quad \rightarrow \quad 2 \ H_2O \quad + \quad O_2$$

[III] As the graph shown in **Figure 3.3** was simulated, all the data points fit nicely on the curve. The corresponding values from a real experiment, however, would be most unlikely to do so!

the rate equation tells us that the reaction does not involve their direct collision in a single reaction step (in which case, the reaction would be second order). The proposed mechanism must proceed in at least two steps, with a single H_2O_2 molecule involved in the slowest, rate-determining step.

It is not unusual to be asked in an examination question to suggest a mechanism for a reaction with which you may not be familiar. Notice my choice of words: 'a mechanism'. You are not expected necessarily to come up with the *correct* mechanism (which may, in fact, not be agreed between chemists), but with a mechanism that is consistent with the rate equation. The more reaction mechanisms you are exposed to in your studies, the more likely you will be able to apply the underlying principles to an unfamiliar reaction, which is why the decomposition of hydrogen peroxide will be looked at in some detail.

As we have seen, in the *absence* of a catalyst the decomposition of hydrogen peroxide begins with the homolytic fission of the O—O single bond, forming two hydroxyl radicals:

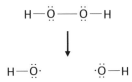

This is the rate-determining step: it is what you are measuring when you measure the rate of H_2O_2 decomposition. Before an individual hydrogen peroxide molecule breaks into a pair of ·OH radicals, it acquires energy by colliding with other molecules (both H_2O_2 and H_2O), forming an 'energised' species we can represent as $(H_2O_2)^*$. It is in this energised state that fission of the O—O bond occurs. As the formation of $(H_2O_2)^*$ is much faster than the bond fission, the overall process displays first-order behaviour (in accordance with the Lindemann model, described above for the popping open of cyclopropane). The subsequent reactions of the two ·OH radicals with the *second* H_2O_2 molecule are so fast, they make no contribution to the observed rate of the reaction.

From your Year-12 studies on the halogenation of alkanes, you will know that halogen radicals (such as ·Cl, generated *via* the homolytic fission of Cl_2 molecules using the energy from ultraviolet light) react with methane by 'pinching off' hydrogen atoms in a chain propagation reaction:

$$CH_4 + \cdot Cl \rightarrow \cdot CH_3 + HCl$$

A very similar reaction takes place between the two hydroxyl radicals and our second molecule of hydrogen peroxide. To help you understand the electron movements that occur, dot-and-cross diagrams are used:

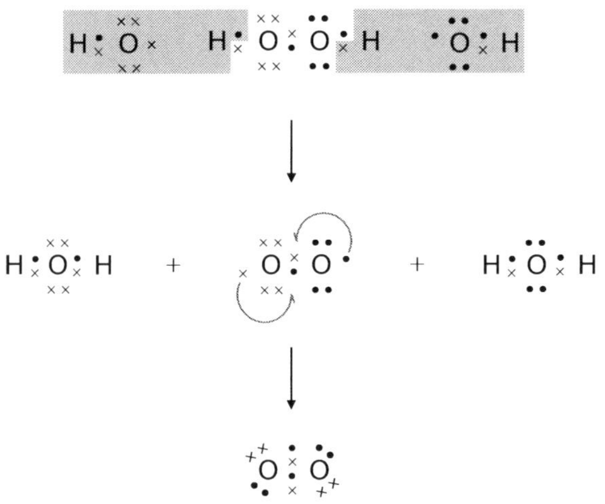

Notice how single-headed ('fish hook') arrows have been used to show the two electrons – one from each oxygen – pairing up to form the second covalent bond in the O_2 molecule. The use of fish-hook arrows, to show the movement of *single* electrons, is not required at A-level, but many teachers use them all the same when showing the mechanisms of reactions involving radicals. The mechanism shown above is a somewhat simplified version of the actual mechanism, in which the H_2O_2 molecule is first attacked by just one hydroxyl radical; the resultant hydroperoxyl radical (HOO·) is then attacked by a second hydroxyl radical (before which it may release a H^+ ion). The reaction is further complicated by the self-reaction of hydroperoxyl radicals,

$$2\,HOO\cdot \rightarrow H_2O + O_2$$

but we need not be concerned with these details. For our purposes, we need only to propose a mechanism that is consistent with the rate equation. We can write our mechanism as follows:

$$H_2O_2 \rightarrow 2 \cdot OH \quad \text{(slow, rate-determining step)}$$

$$2 \cdot OH + H_2O_2 \rightarrow 2 H_2O + O_2 \quad \text{(fast step)}$$

As stated earlier, the very slow decomposition of hydrogen peroxide that occurs in solutions under normal laboratory conditions of temperature and pressure (or in a refrigerator) is invariably due to catalysis by contaminating metal ions. Most laboratory solutions contain Fe^{3+} ions at trace levels (below 1×10^{-6} mol dm^{-3}), which is enough for the reaction to proceed. We can envisage the catalytic cycle occurring in two steps, which when added together cancel down to give the equation for the disproportionation of the peroxide:

$$2\,\cancel{Fe}^{3+} + H_2O_2 \rightarrow 2\,\cancel{Fe}^{2+} + O_2 + 2\,\cancel{H}^+$$

$$2\,\cancel{Fe}^{2+} + H_2O_2 + 2\,\cancel{H}^+ \rightarrow 2\,\cancel{Fe}^{3+} + 2 H_2O$$

$$\overline{2 H_2O_2 \rightarrow 2 H_2O + O_2}$$

You will not be surprised to learn that both steps are somewhat more complicated than shown in these equations. The initial reduction of Fe^{3+} by H_2O_2, the slower of the two reactions, involves the transfer of a single electron from an O−H bond in the peroxide, leaving the hydroperoxyl radical (HOO$\cdot$) and a proton:

135

The hydroperoxyl radical then transfers an electron to a second Fe^{3+} ion, giving an O_2 molecule and another proton:

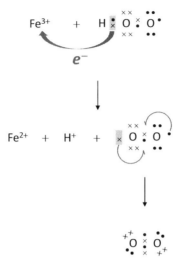

Thus, the two reaction steps in the reduction of Fe^{3+} by hydrogen peroxide are:

$$Fe^{3+} + H_2O_2 \rightarrow Fe^{2+} + HOO^{\bullet} + H^+$$

$$Fe^{3+} + HOO^{\bullet} \rightarrow Fe^{2+} + O_2 + H^+$$

The oxidation of the Fe^{2+} back to Fe^{3+} by the second molecule of hydrogen peroxide, which is the much faster reaction, also occurs in two steps. In the initial step, known as the Fenton reaction, the hydroxyl radical is formed:

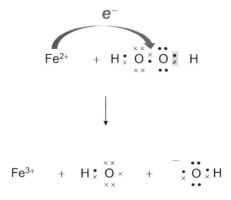

The hydroxyl radical, an extremely powerful oxidant, then removes an electron from the second Fe^{2+} ion. The two reaction steps are summarised as follows:

$$Fe^{2+} + H_2O_2 \rightarrow Fe^{3+} + HO^{\bullet} + OH^-$$

$$Fe^{2+} + HO^{\bullet} \rightarrow Fe^{3+} + OH^-$$

The two hydroxide ions formed in these reactions combine with the two hydrogen ions that were formed during the reduction of Fe^{3+} by hydrogen peroxide, forming two molecules of water. With appropriate cancelling out, it may be seen that Fe^{3+} ions act catalytically in their promotion of hydrogen peroxide disproportionation:

$$Fe^{3+} + H_2O_2 \rightarrow Fe^{2+} + HOO^{\bullet} + H^+$$

$$Fe^{3+} + HOO^{\bullet} \rightarrow Fe^{2+} + O_2 + H^+$$

$$Fe^{2+} + H_2O_2 \rightarrow Fe^{3+} + HO^{\bullet} + OH^-$$

$$Fe^{2+} + HO^{\bullet} \rightarrow Fe^{3+} + OH^-$$

$$2H^+ + 2OH^- \rightarrow 2H_2O$$

_____

$$2H_2O_2 \rightarrow 2H_2O + O_2$$

CHAPTER 3 SUMMARY – the key points

- One of the reasons for investigating the kinetics of a chemical reaction is to gain clues to its mechanism: the rate equation reveals the species involved in the reaction's rate-determining step (RDS).

- Although chemists may be able to suggest a possible mechanism for a given reaction, based on the known properties of the reactant molecules, any mechanism must be supported by kinetic data. The correct procedure is to obtain the rate equation first and *then* propose a mechanism with which it is compatible.

- A reaction that occurs in more than one step has a transition state for each reaction step. The transition state with the highest activation energy occurs in the RDS. Whereas **transition states** occur at energy maxima, **reaction intermediates** occur in the 'wells' at energy minima. Intermediates are generated in one step (elementary reaction) of a multistep reaction mechanism and are then consumed in a subsequent step, so they do not appear as products.

- Whereas primary halogenoalkanes undergo mainly second order nucleophilic substitutions, tertiary halogenoalkanes react mainly through a first-order process (involving their formation of a carbocation intermediate). Nucleophilic substitution in tertiary halogenoalkanes by the hydroxide ion is in competition with an elimination reaction, which generates an alkene. Reaction conditions can be selected to favour the reaction of hydroxide ions with tertiary halogenoalkanes by substitution or elimination, the latter being favoured at high temperature and the use of ethanolic alkali.

- Before iodine can react with propanone, the ketone must first isomerise into an enol, which is the RDS. Hydrogen ions (H^+) catalyse the isomerisation, so appear in the rate equation:

$$\text{rate} \;=\; k\,[H^+][CH_3COCH_3]$$

Iodine, which reacts rapidly with the enol, does not appear in the rate equation: its rate of reaction is limited by the availability of the enol, formed in the slow step.

- Iodide ions are oxidised by hydrogen peroxide in the Harcourt-Essen reaction:

$$H_2O_2 \;+\; 2\,I^- \;+\; 2\,H^+ \;\rightarrow\; 2\,H_2O \;+\; I_2$$

The reaction proceeds simultaneously by two, slightly different mechanisms – one being second order, the other third order. The respective rate equations are:

$$\text{rate} \;=\; k_2\,[H_2O_2][I^-]$$

$$\text{rate} \;=\; k_3\,[H_2O_2][I^-][H^+]$$

Although the rate constant of third-order reaction (k_3) is the greater, this mechanism assumes importance only at relatively high hydrogen ion concentrations. At low concentrations of H^+, the observed kinetics are second order, with changes in $[H^+]$ having no effect on the observed rate.

In an examination question, it is down to you to deduce from the data provided which of the two rate equations applies: if the data show the reaction to be zero order with respect to the H^+ ion, then you would state that the reaction is second order and give the appropriate rate equation. It is 'unlikely' you would be expected to disentangle the two mechanisms: you are more likely to be given data that point decisively to one of the two rate equations.

- Hydrogen peroxide can be replaced by the peroxodisulfate(VI) ion as the oxidising agent:

$$S_2O_8^{2-} \; + \; 2\,I^- \; \rightarrow \; 2\,SO_4^{2-} \; + \; I_2$$

This reaction has the following, second-order rate equation,

$$\text{rate} \; = \; k\,[S_2O_8^{2-}][I^-]$$

and is catalysed by Fe^{2+} ions:

$$S_2O_8^{2-} \; + \; 2\,Fe^{2+} \; \rightarrow \; 2\,SO_4^{2-} \; + \; 2\,Fe^{3+}$$

$$2\,Fe^{3+} \; + \; 2\,I^- \; \rightarrow \; I_2 + \; 2\,Fe^{2+}$$

- Iron ions also catalyse the decomposition of hydrogen peroxide, which we can summarise in the reactions:

$$2\,Fe^{3+} \; + \; H_2O_2 \; \rightarrow \; 2\,Fe^{2+} \; + \; O_2 \; + \; 2\,H^+$$

$$2\,Fe^{2+} \; + \; H_2O_2 \; + \; 2\,H^+ \; \rightarrow \; 2\,Fe^{3+} \; + \; 2\,H_2O$$

$$2\,H_2O_2 \; \rightarrow \; 2\,H_2O \; + \; O_2$$

In the absence of a catalyst, hydrogen peroxide undergoes very slow decomposition. Even though the overall reaction involves two H_2O_2 molecules,

$$2 H_2O_2 \rightarrow 2 H_2O + O_2$$

the reaction is first order, with the rate equation:

$$\text{rate} = k [H_2O_2]$$

- Although the rate equation for the uncatalysed decomposition of hydrogen peroxide would appear to preclude the possibility that the reaction involves a collision between two H_2O_2 molecules, its first-order behaviour can be explained using the Lindemann model. According to this theory, which was explained using the isomerisation of cyclopropane to propene as an example, first-order reactions do involve collisions between pairs of molecules, but do not display second-order kinetics because the RDS is the transition of a *single*, energetically-excited reactant molecule through the transition state to form the products.

Chapter 4

KINETICS IN THE SCHOOL LABORATORY:

METHODS OF INVESTIGATION

In the preceding chapters, the methods used to measure the rates of the reactions described have barely been mentioned – at best, only in passing, and in some cases not at all. Indeed, the focus of our interests thus far has been the 'fruits' – or outcomes – of laboratory investigations: namely rate equations, rate constants and the underlying chemical mechanisms they reflect. We now turn our attention to the experimental methods used in reaction kinetics. As it is still fresh in our minds from Chapter 3, we will begin by looking at methods used to investigate the kinetics of the Harcourt-Essen reaction.

4.1   Revisiting the Harcourt-Essen reaction: redox titration

In planning to measure the rate of any chemical reaction, it is necessary to identify an observable change that can be monitored over time. This may be a colour change, the evolution of a gas or a change in pH. You should recall from your studies of the halogens (Year 12) that aqueous solutions of iodine are brown. As aqueous solutions of the iodide ion are colourless, we might simply measure the rate of the Harcourt-Essen reaction by determining the rate at which the reaction mixture turns brown with the formation of iodine:

$$H_2O_2 \quad + \quad 2\,I^- \quad + \quad 2\,H^+ \quad \rightarrow \quad 2\,H_2O \quad + \quad I_2$$

Reactions that result in a change in colour can be monitored using a simple colorimeter, fitted with an appropriate filter. These instruments are available in some school laboratories. The brown colouration that develops during the course of the Harcourt-Essen reaction, however, is not due to iodine itself; rather it is from the tri-iodide ion (I_3^-), which is formed in the reaction between iodine and the iodide ion in the reversible reaction:

$$I^- \quad + \quad I_2 \quad \rightleftharpoons \quad I_3^-$$

As the concentration of the triiodide ion is determined by the concentrations of both iodine and the iodide ion, there is no direct, linear relationship between the intensity of the brown colour and the progress of the reaction. The concentration of I_3^- at any moment reflects the establishment of an equilibrium between all three species.[mmm] Think of it like this: during the course of the reaction, iodine is generated, which increases the concentration of I_3^- (the equilibrium moves to the right), but *at the same time* iodide is being depleted, resulting in a decrease in the concentration of I_3^- (the equilibrium moves to the left).

We owe a great deal of our understanding of the kinetics of the Harcourt-Essen reaction to the pioneering work carried out by Herman Liebhafsky and Ali Mohammad at the University of California in the early 1930s. These researchers measured the rate of iodine formation by titration against the thiosulfate ion. Although you are unlikely to use this method to investigate the rate of the Harcourt-Essen reaction in your school practical work, you will almost certainly carry out iodine-thiosulfate titrations to determine the concentrations of various oxidants (typically Cu^{2+} and ClO^- ions) in your wider A-level laboratory work.[nnn]

It is helpful to think of the thiosulfate ion ($S_2O_3^{2-}$) as a sulfate ion (SO_4^{2-}) in which one of the oxygen atoms has been replaced by a second sulfur atom. The thiosulfate ion is a powerful reducing agent. When reducing an iodine molecule to two iodide ions,

$$I_2 \quad + \quad 2\,e^- \quad \rightarrow \quad 2\,I^-$$

two thiosulfate ions are involved, forming the tetrathionate ion ($S_2O_6^{2-}$):

[mmm] An equilibrium can be established only in a closed system, *i.e.* a system to which no energy (heat), reactant or product is being added or removed. Although I_2 is being added (generated) and I^- removed during the Harcourt-Essen reaction, the rate of the reaction is extremely slow compared with the rate at which these species establish an equilibrium with I_3^-. This means that, although the concentrations of all three species are changing continuously, there is, at any 'instant' during the reaction, sufficient time for them to reach equilibrium (giving the concentrations prescribed by the equilibrium constant, K_c).

[nnn] Redox titrations, including the $I_2/S_2O_3^{2-}$ system, are included in Section 5.2.3 ('Redox and electrode potentials') of the OCR specification. Section 3.2.5.5 ('Variable oxidation states') of the AQA specification refers specifically to MnO_4^- redox titrations, but then states that students should be able to perform calculations for 'similar' redox reactions. The $I_2/S_2O_3^{2-}$ system is covered in Topic 14 (Redox II) of the Edexcel specification.

$$2\,S_2O_3^{2-} \quad \rightarrow \quad S_4O_6^{2-} + \quad 2\,e^-$$

Combining these half-equations, and cancelling out the electrons, gives the overall reaction:

$$I_2 + 2\,S_2O_3^{2-} \quad \rightarrow \quad 2\,I^- + \quad S_4O_6^{2-}$$

You are unlikely to be questioned on the mechanistic details of this reaction at A-level, but it is important you have a 'feel' for what is going on, including an awareness of the structure of the thiosulfate and tetrathionate ions. In saying that, the examiners are forever seeking out novel reaction systems in which to probe the ability of candidates to apply, in unfamiliar situations, the core principles they have been taught. One sees many examination questions that appear to be on topics that are not covered in the specifications, but in fact this is rarely the case. The more experience you can acquire in the application of the taught material to unfamiliar reactions, the better prepared you will be to handle such questions. The oxidation of thiosulfate to tetrathionate could, for example, be used to probe your understanding of oxidation states (oxidation numbers).

A somewhat simplified version of the mechanism for the reduction of iodine to iodide ions is given below, where it can be seen that the two electrons needed to reduce the former are donated by the first thiosulfate ion.

Liebhafsky and Mohammad describe how, in a typical experiment, they removed 25 cm³ samples (aliquots) from the reaction mixture at selected time-points and titrated these immediately against 0.002 mol dm⁻³ thiosulfate.[ooo] The moment it was removed from the reaction mixture, each aliquot was added to 125 cm³ of ice-cold water: the resultant decrease in temperature and dilution of the reactants retards the reaction to the extent that it can be considered to have been stopped. In one such series of experiments, the initial concentrations of H_2O_2 (added last, to start the reaction) and I^- ions were held at 8.00×10^{-4} and 5.72×10^{-2} mol dm⁻³, respectively, with the initial concentration of H^+ being varied over a wide range.

Plots **(a)**, **(b)** and **(c)** in **Figure 4.1** show the time-courses of iodine generation at H^+ concentrations of 2.10×10^{-2}, 4.20×10^{-2} and 2.52×10^{-1} mol dm⁻³, respectively.[PPP] Notice how the units of iodine concentration given on the y-axis have been multiplied by 10^4 before being plotted. Such 'scaling' of the units avoids the cumbersome use of numbers in standard form on the axis and is therefore common practice. Increasing the concentration of H^+ ions clearly increases the rate. Indeed, at the highest concentration of H^+ shown (0.252 mol dm⁻³), iodine formation ceases after about 25 minutes, after reaching a concentration of approximately 8.00×10^{-4} mol dm⁻³. This is because the limiting reactant, H_2O_2, is present at a concentration of 8.00×10^{-4} mol dm⁻³ and thereby determines the maximum amount of iodine that can be generated, as reflected in the stoichiometry for the overall reaction (for each mole of H_2O_2 used, one mole of I_2 is generated):

$$H_2O_2 \ + \ 2\,I^- \ + \ 2\,H^+ \quad \rightarrow \quad 2\,H_2O \ + \ I_2$$

[ooo] H. A. Liebhafsky and A. Mohammad (1933). The kinetics of the reduction, in acid solution, of hydrogen peroxide by iodide ion. *J. Am. Chem. Soc.* **55**, 3977 – 3986. Although the authors do not appear to specify the chemical source of the thiosulfate ion they used, it is usually added as sodium thiosulfate ($Na_2S_2O_3$).
[PPP] Although the reactant concentrations given in **Figure 4.1** are based on those given in the Liebhafsky and Mohammad paper (see Footnote[ooo]), the graphs in the paper do not show changes in iodine concentration; rather, the authors plotted the titre of thiosulfate solution required at each time-point. In the interests of simplicity and clarity, however, iodine concentrations have been plotted in the graphs presented here.

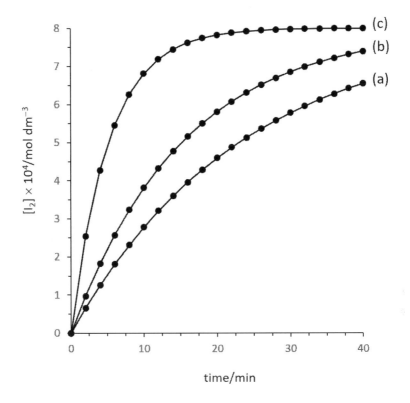

Figure 4.1   Effect of the initial H^+ ion concentration on the rate of iodine formation in the Harcourt-Essen reaction. The initial concentrations of H_2O_2 and the iodide ion were 8.00×10^{-4} and 5.72×10^{-2} mol dm^{-3}, respectively. The initial concentrations of H^+ in plots **(a)**, **(b)** and **(c)** were 2.10×10^{-2}, 4.20×10^{-2} and 2.52×10^{-1} mol dm^{-3}, respectively.

By drawing a tangent to each curve at time zero, it is possible to calculate the initial rate of the reaction at each concentration of the H^+ ion. **Figure 4.2** shows a tangent drawn to curve **(b)**, in which the initial concentration of H^+ was 4.20×10^{-2} mol dm^{-3}. The figure also shows where the measurements used to calculate the gradient of the tangent were taken.

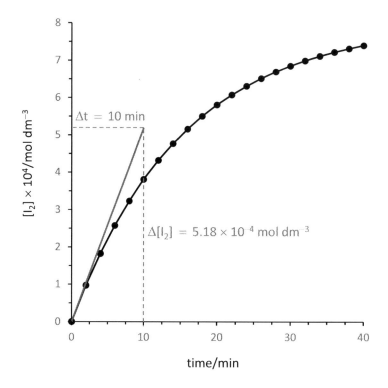

Figure 4.2   Using a tangent drawn to the curve at the zero timepoint to obtain the initial rate of the Harcourt-Essen reaction. The initial reactant concentrations are those given for curve (b) in **Figure 4.1.** See text for details.

Dividing the change in iodine concentration, $\Delta[I_2]$ (where the symbol Δ, pronounced 'delta', signifies a change in something) by Δt gives us a rate of 5.18×10^{-5} mol dm^{-3} min^{-1}:

$$\text{rate} = \frac{5.18 \times 10^{-4} \ \text{mol dm}^{-3}}{10 \ \text{min}}$$

$$= 5.18 \times 10^{-5} \ \frac{\text{mol dm}^{-3}}{\text{min}}$$

$$= 5.18 \times 10^{-5} \ \text{mol dm}^{-3} \text{min}^{-1}$$

146

The initial rates obtained by drawing tangents to all three curves in **Figure 4.1** are given in **Table 4.1**, from which it is evident that the reaction is not first-order with respect to the H^+ ion. Thus, increasing the concentration of H^+ two-fold, from 0.0210 to 0.0420 mol dm^{-3} causes the rate to increase by less than 25 %. Had the reaction been first-order in $[H^+]$, the rate would have doubled. Similarly, increasing $[H^+]$ 6-fold, from 0.0420 to 0.252 mol dm^{-3} results in only a 3-fold increase in the rate.

$[H^+]$/ mol dm^{-3}	initial rate/ mol dm^{-3} min^{-1}
0.0210	4.17×10^{-5}
0.0420	5.18×10^{-5}
0.2520	1.53×10^{-4}

Table 4.1   Effect of H^+ concentration on the rate of the Harcourt-Essen reaction. The initial rates were obtained from tangents drawn at the zero-time point of the plots shown in **Figure 4.1**, where further experimental details are given.

The observed response of the rate to changes in H^+ concentration reflects, of course, the fact that the reaction proceeds by two separate mechanisms, as described in Chapter 3 and encapsulated in the overall rate equation:

$$\text{rate} \quad = \quad k_2 [H_2O_2][I^-] \quad + \quad k_3 [H_2O_2][I^-][H^+]$$

One mechanism (with the rate constant k_2) is zero-order in H^+, but the other (with the rate constant k_3) is first-order in H^+. The 'observed' order of the reaction with respect to the H^+ ion depends on the extent to which each mechanism is contributing to the overall rate under a particular set of reactant concentrations. As we saw in Chapter 3, the reaction is observed to be zero-order in H^+ only when the concentration of the ion is so low that the mechanism controlled by k_3 is making only a negligible contribution to the overall rate.

We can obtain the values of k_2 and k_3 by substituting into the rate equation the data from any **two** experiments. Using, for example, the first two sets of data from **Table 4.1** (where $[H^+]$ is 0.0210 and 0.0420 mol dm^{-3}, respectively), we have,

$$4.17 \times 10^{-5}\,\text{mol dm}^{-3}\,\text{min}^{-1} =$$

$$k_2(8.00 \times 10^{-4}\,\text{mol dm}^{-3})(5.72 \times 10^{-2}\,\text{mol dm}^{-3}) +$$

$$k_3(8.00 \times 10^{-4}\,\text{mol dm}^{-3})(5.72 \times 10^{-2}\,\text{mol dm}^{-3})(\mathbf{2.10 \times 10^{-2}\,\text{mol dm}^{-3}})$$

and,

$$5.18 \times 10^{-5}\,\text{mol dm}^{-3}\,\text{min}^{-1} =$$

$$k_2(8.00 \times 10^{-4}\,\text{mol dm}^{-3})(5.72 \times 10^{-2}\,\text{mol dm}^{-3}) +$$

$$k_3(8.00 \times 10^{-4}\,\text{mol dm}^{-3})(5.72 \times 10^{-2}\,\text{mol dm}^{-3})(\mathbf{4.20 \times 10^{-2}\,\text{mol dm}^{-3}})$$

which multiplying-out (including the units[qqq]) give,[rrr]

$$4.17 \times 10^{-5}\,\text{mol dm}^{-3}\,\text{min}^{-1} =$$

$$k_2(4.576 \times 10^{-5}\,\text{mol}^2\,\text{dm}^{-6}) + k_3(9.610 \times 10^{-7}\,\text{mol}^3\,\text{dm}^{-9})$$

and,

$$5.18 \times 10^{-5}\,\text{mol dm}^{-3}\,\text{min}^{-1} =$$

$$k_2(4.576 \times 10^{-5}\,\text{mol}^2\,\text{dm}^{-6}) + k_3(1.922 \times 10^{-6}\,\text{mol}^3\,\text{dm}^{-9})$$

Subtracting the first equation from the second removes the k_2 term, allowing k_3 to be found:[sss]

[qqq] Exponents ('powers') applied to numbers *and* units are added during multiplication. Thus, 8.00×10^{-4} mol dm^{-3} multiplied by 5.72×10^{-2} mol dm^{-3} is 4.58×10^{-5} mol^2 dm^{-6} (mol^1 dm^{-3} × mol^1 dm^{-3} = mol^2 dm^{-6}).

[rrr] Notice how 'too many' significant figures have been given after the multiplication step. Rounding down to 3 s.f. at this stage runs the risk of introducing a rounding error in the final answer: always round down to the appropriate number of significant figures only in the final step.

[sss] Note how the exponents in the units are *subtracted* from each other during division. Thus, dm^{-3} divided by dm^{-9} is dm^6 ($-3 - -9 = -3 + 9 = +6$) *etc.*

$$1.01 \times 10^{-5}\,\text{mol dm}^{-3}\,\text{min}^{-1} \;=\; k_3\,(9.610 \times 10^{-7}\,\text{mol}^3\,\text{dm}^{-9})$$

$$k_3 \;=\; \frac{1.01 \times 10^{-5}\,\text{mol dm}^{-3}\,\text{min}^{-1}}{9.610 \times 10^{-7}\,\text{mol}^3\,\text{dm}^{-9}}$$

$$=\; 10.5\;\frac{\text{mol dm}^{-3}\,\text{min}^{-1}}{\text{mol}^3\,\text{dm}^{-9}}$$

$$=\; 10.5\,\text{dm}^6\,\text{mol}^{-2}\,\text{min}^{-1}$$

Converting the units of time from minutes to seconds gives the value of k_3 reported in Chapter 3 (0.175 $\text{dm}^6\,\text{mol}^{-2}\,\text{s}^{-1}$). It is now a simple matter to substitute the value of k_3 back into the rate equation, along with a set of concentrations and the corresponding rate, to obtain k_2. Thus, taking again the first set of values given in **Table 4.1** we have,

$$4.17 \times 10^{-5}\,\text{mol dm}^{-3}\,\text{min}^{-1} \;=$$

$$k_2\,(8.00 \times 10^{-4}\,\text{mol dm}^{-3})(5.72 \times 10^{-2}\,\text{mol dm}^{-3}) \;+$$

$$(10.5\,\text{dm}^6\,\text{mol}^{-2}\,\text{min}^{-1})(8.00 \times 10^{-4}\,\text{mol dm}^{-3})$$

$$(5.72 \times 10^{-2}\,\text{mol dm}^{-3})(2.10 \times 10^{-2}\,\text{mol dm}^{-3})$$

thus,

$$4.17 \times 10^{-5}\,\text{mol dm}^{-3}\,\text{min}^{-1} \;=$$

$$k_2\,(4.576 \times 10^{-5}\,\text{mol}^2\,\text{dm}^{-6}) \;+\; (1.009 \times 10^{-5}\,\text{mol dm}^{-3}\,\text{min}^{-1})$$

giving,

$$3.161 \times 10^{-5}\,\text{mol dm}^{-3}\,\text{min}^{-1} \;=\; k_2\,(4.576 \times 10^{-5}\,\text{mol}^2\,\text{dm}^{-6})$$

$$k_2 \;=\; 0.691\,\text{dm}^3\,\text{min}^{-1}\,\text{mol}^{-1}$$

$$\text{or}\;\; 0.0115\,\text{dm}^3\,\text{mol}^{-1}\,\text{s}^{-1}$$

Liebhafsky and Mohammad obtained the values of k_2 and k_3 not by solving a pair of simultaneous equations, as shown above, but by taking a more sophisticated approach that involved using the reaction rates measured at several H^+ concentrations, covering a wide range of values.

This approach relies on the fact that, at the reactant concentrations used by the authors (and here), the reaction is essentially first-order in H_2O_2 and zero-order in both H^+ and I^-. We have seen already that hydrogen peroxide is the limiting reactant. Now consider the concentrations of I^- and H^+ ions at the end of the reaction, when all the H_2O_2 has reacted: this occurs after approximately 25 min in the reaction plotted as curve (c) in **Figure 4.1**. The initial concentration of I^- was 0.0572 mol dm^{-3}. In the interests of simplicity, we will take the total volume of the reaction mixture to be 1 dm^3, which means there are 0.0572 moles of iodide at the start of the reaction (at the zero time-point):

$$H_2O_2 \quad + \quad 2\,I^- \quad + \quad 2\,H^+ \quad \rightarrow \quad 2\,H_2O \quad + \quad I_2$$

initial number of moles in each dm^3	0.0008	0.0572	0.252	55.6^{ttt}	0
final number of moles in each dm^3	0	0.0556	0.250	55.6	0.008

Since *two* moles of I^- react with each mole of H_2O_2, at the end of the reaction 0.0016 moles of I^- will have reacted, leaving 0.0556 moles, giving a final I^- concentration of 0.0556 mol dm^{-3}. Thus, while the concentration of H_2O_2 has fallen to zero, that of the iodide ion has decreased by less than 3 %. Similarly, it is seen that the concentration of H^+ ions falls by only 0.8 % over the course of the reaction. As their

---

[ttt] As we saw earlier (page 35), the concentration of H_2O in pure water is obtained by dividing the mas of 1 dm^3 of water (1000 g) by the molar mass of H_2O (18.0 g mol^{-1}). The contribution of the water molecules generated in the reaction to the final concentration of H_2O is, of course, negligible.

concentrations remain essentially constant, I^- and H^+ cannot be causing the reaction rate to change over the course of the reaction: to all intents and purposes, the reaction of zero-order in $[I^-]$ and $[H^+]$. The rate equation can now be considered to contain only one independent variable on the right-hand side, namely $[H_2O_2]$:

$$\text{rate} \quad = \quad k_2[H_2O_2][I^-] \quad + \quad k_3[H_2O_2][I^-][H^+]$$

Gathering together all the constants on the right-hand side (the two rate constants, $[I^-]$ and $[H^+]$) into a single term puts the rate equation in the form,

$$\text{rate} \quad = \quad [I^-](k_2 + k_3[H^+]) \quad \times \quad [H_2O_2]$$

As long as I^- and H^+ are present at excess concentration over H_2O_2, the term $[I^-](k_2 + k_3[H^+])$ is a constant. Taking the concentrations used in the experiment plotted as curve (c) in **Figure 4.1**, we have,

$$[I^-](k_2 + k_3[H^+]) \quad = \quad 0.0572 \text{ mol dm}^{-3}(k_2 + k_3 0.252 \text{ mol dm}^{-3})$$

which, using the values of k_2 and k_3 given above, becomes:[uuu]

$$[I^-](k_2 + k_3[H^+]) \quad =$$

$$(0.0572 \text{ mol dm}^{-3})(0.691 \text{ dm}^3 \text{ mol}^{-1} \text{ min}^{-1}) \; +$$

$$(0.0572 \text{ mol dm}^{-3})(10.5 \text{ dm}^6 \text{ mol}^{-2} \text{ min}^{-1})(0.252 \text{ mol dm}^{-3})$$

$$= \quad 0.03952 \text{ min}^{-1} \quad + \quad 0.1513 \text{ min}^{-1}$$

$$= \quad 0.191 \text{ min}^{-1}$$

The rate equation,

$$\text{rate} \quad = \quad [I^-](k_2 + k_3[H^+]) \quad \times \quad [H_2O_2]$$

may now be written as,

$$\text{rate} \quad = \quad 0.191 \text{ min}^{-1} \quad \times \quad [H_2O_2]$$

[uuu] Whilst I have every confidence students of Λ-level chemistry are capable of obtaining the correct numerical answer from this calculation, it is equally important they are able to derive the correct units, which is why the calculation is shown in full.

The constant, 0.191 min^{-1}, is referred to as a **pseudo first-order rate constant** and is given the symbol k': it is a first-order rate constant that has been artificially 'imposed' on the reaction by ensuring that I^- and H^+ are present at excess concentration over H_2O_2. As the value of k' depends on the particular set of I^- and H^+ concentrations used,

$$k' = [I^-](k_2 + k_3[H^+])$$

k' is not a constant in the sense that it is valid at all reactant concentrations (as are k_2 and k_3).

Physical chemists often carry out reactions under pseudo first-order conditions because it simplifies the analysis of the reaction kinetics, making it much easier to, for example, determine the reaction order with respect to the concentration of each reactant and to extract the underlying rate constants. Indeed, you will often encounter reactions described in examination questions in which all the reactants except one are present at excess concentration.

Although they may not use such phrases as 'pseudo first-order conditions', the examiners may well ask you why such reactions are zero-order with the respect to the reactants that are present at excess concentration. Similarly, any rate constants you may be expected to obtain from data under such conditions will, in fact, be pseudo first-order rate constants. Thus, while pseudo first-order rate constants may not be referred to as such in the A-level specifications, it is as well that you are familiar with the concept and are able to recognise them for what they are.

Let's now look at how Liebhafsky and Mohammad used pseudo first-order rate constants to obtain k_2 and k_3. Cross-multiplication (*i.e.* dividing both sides by $[I^-]$) and rearranging the expression given above for k' converts it into the form of an equation for a straight line ($y = mx + c$):

$$\frac{k'}{[I^-]} = k_2 + k_3[H^+]$$

$$\frac{k'}{[I^-]} = k_3[H^+] + k_2$$

$$y = mx + c$$

Plotting $k'/[I^-]$ on the y-axis against $[H^+]$ on the x-axis should give a straight line with a gradient (m) equal to k_3 and an intercept on the on the y-axis (c) equal to k_2. **Table 4.2** gives the values of $k'/[I^-]$ calculated when was k' measured at selected H^+ concentrations (taken from the wide range of $[H^+]$ values used by Liebhafsky and Mohammad in their 1933 study, cited in the Footnote[ooo] on page 144). In each case, the initial concentration of the iodide ion, $[I^-]$, was 0.0572 mol dm^{-3}.

As shown in **Figure 4.3**, plotting $k'/[I^-]$ against $[H^+]$ does, indeed, give a straight line, the gradient of which (m) is equal to k_3, calculated as follows:

$$k_3 = \frac{5.25 \text{ mol dm}^{-3} \text{ min}^{-1}}{50.0 \times 10^{-2} \text{ mol dm}^{-3}}$$

$$= 10.5 \text{ min}^{-1}$$

$[H^+] \times 10^2$ /mol dm^{-3}	k' /min^{-1}	$\dfrac{k'}{[I^-]}$ /dm^3 min^{-1} mol^{-1}
4.20	0.0646	1.13
8.34	0.0898	1.57
16.8	0.140	2.45
25.2	0.190	3.32
42.0	0.292	5.10
55.0	0.370	6.47

Table 4.2 Value of the pseudo first-order rate constant (k') for the Harcourt-Essen reaction measured over a range of initial concentrations of the H^+ ion. The initial concentrations of H_2O_2 and the iodide ion were 8.00×10^{-4} and 5.72×10^{-2} mol dm^{-3}, respectively. The values given in the third column were obtained by dividing k' by the concentration of the iodide ion, prior to plotting against $[H^+]$ (**Figure 4.3**). See text for details.

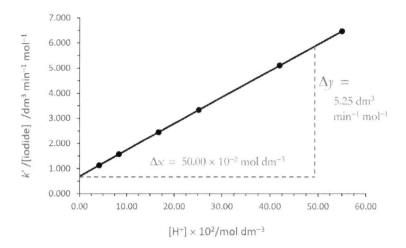

Figure 4.3 Plot of the $k'/[I^-]$ and $[H^+]$ values reported in **Table 4.2**, confirming the data can be 'fitted' to an equation of the form $y = mx + c$. The figure also shows the measurements used to calculate the gradient of the straight line, which is equal to k_3 (see text for details).

When the concentration H^+ is zero, our equation simplifies, such that the term $k'/[I^-]$ is equal to k_2:

$$\frac{k'}{[I^-]} = k_3[H^+] + k_2$$

$$\frac{k'}{[I^-]} = k_2 \times 0 + k_2$$

$$\frac{k'}{[I^-]} = k_2$$

Extending the line-of-best-fit back to 0.00 $[H^+]$, where it intercepts the y-axis, puts k_2 at 0.691 dm^3 min^{-1} mol^{-1} (*i.e.* 0.0115 dm^3 mol^{-1} s^{-1}), in agreement with the value obtained earlier by solving a pair of simultaneous equations. Although the current method is more time consuming, it is to be preferred because it takes in to account the value

154

of k' obtained at several H^+ concentrations, so any anomalies in the data are less likely to adversely affect the values of k_2 and k_3 obtained.

4.2   The iodine-clock method

In addition to the redox-titration method described above, the Harcourt-Essen reaction lends itself to investigation using a 'clock' method. Indeed, due to its simplicity, the so-called 'iodine clock' is invariably the method of choice when studying this reaction in the school laboratory. A clock method is any method in which the rate of a reaction is measured as the time taken for a fixed amount of product to be formed. A very simple example, which you may have encountered in your GCSE studies, is the 'disappearing cross' method, used to measure the rate of the reaction between hydrochloric acid and sodium thiosulfate (typically at several different temperatures):

$$Na_2S_2O_{3\ (aq)} \ + \ 2\,HCl_{(aq)} \ \rightarrow$$

$$2\,NaCl_{(aq)} \ + \ SO_{2\ (g)} \ + \ S_{(s)} \ + \ H_2O_{(l)}$$

The reaction is carried out in a beaker placed over a piece of paper with a large 'X' written on it. Viewed from above, the X is initially visible but, as the reaction proceeds, the sulfur that is produced causes the mixture to become cloudy. The rate of the reaction is inversely proportional to the time taken for the X to no longer be visible through the reaction mixture:

$$rate \ \propto \ \frac{1}{time}$$

The shorter the time taken for the cross to 'disappear', the higher the rate. If we wished to carry out a detailed kinetic analysis of the reaction we might, for example, weigh the amount of sulfur needed to block the cross from view. We would then have the time taken for a particular amount of sulfur to be produced, from which it would be possible to express the rate as a change in reactant concentration per unit time.

The iodine-clock method, when used to investigate the Harcourt-Essen reaction, involves carrying out the reaction in the presence of starch and a fixed amount of sodium thiosulfate. Although the iodine generated in the reaction can form the triiodide ion (I_3^-), which enters the helical structure of starch, giving the characteristic deep, blue-black complex,[vvv]

$$H_2O_2 \;+\; 2\,I^- \;+\; 2\,H^+ \;\rightarrow\; 2\,H_2O \;+\; I_2$$

$$I^- \;+\; I_2 \;\rightleftharpoons\; I_3^-$$

$$I_3^- \;+\; starch \;\rightarrow\; blue\text{-}black\ complex$$

thiosulfate ions initially prevent the formation of the complex by rapidly reducing iodine back to iodide ions:

$$I_2 \;+\; 2\,S_2O_3^{2-} \;\rightarrow\; 2\,I^- \;+\; S_4O_6^{2-}$$

It is only when all the all of thiosulfate ions present have been oxidised to tetrathionate that the blue-black complex with starch is formed. This occurs quite suddenly, making it easy for the student to know when to stop the timer.

Since the rate is inversely proportional to the time (t) taken for the solution to turn blue-black, simple plots of the raw data allow the order of the reaction to be determined with respect to each reactant. To measure the order with respect to the iodide ion, for example, a plot of $1/t$ against the initial concentration of the ion will reveal the order. Unlike the approach described above, using the titration method, when using the iodine-clock method it important that *all* the reactants – including the one being varied in concentration – are present at much higher concentration than the thiosulfate ion. This is to ensure that, during the time interval before the blue-black complex appears, the concentrations of the reactants decrease by only a very small amount – effectively remaining unchanged, at their initial values. The best way to explain this is with an example.

Consider a series of reaction runs designed to investigate the effect of changing the concentration of the iodide ion on the rate of the Harcourt-Essen reaction at 25 °C. In each reaction mixture, which has a final volume of 100 cm^3, the initial concentrations of H_2O_2, HNO_3 and $Na_2S_2O_3$ are 8.00×10^{-3}, 0.150 and 2.50×10^{-4} mol dm^{-3}, respectively.[www] The concentration of KI is varied from 5.00×10^{-3} up

---

[vvv] This interaction forms the basis of the test for starch used in biology.
[www] Since nitric acid undergoes complete dissociation in water
$\quad$ ($HNO_3 \rightarrow H^+ + NO_3^-$), the concentration of H^+ will be 0.150 mol dm^{-3}.

to 2.50×10^{-2} mol dm^{-3}. The 100 cm^3 reaction mixture also contains 2.50×10^{-5} moles of the $S_2O_3{}^{2-}$ ion[xxx] and a trace of starch. Since one molecule of iodine reacts with two thiosulfate ions (see equation, above), it follows that the blue-black complex appears the moment 1.25×10^{-5} moles of I_2 have been generated.

As shown below, where the initial concentration of KI has been taken to be 1.50×10^{-2} mol dm^{-3}, the change in the amount of each reactant that occurs with the formation of 1.25×10^{-5} moles of I_2 is readily calculated. The 100 cm^3 mixture initially contains 8.00×10^{-4} moles of H_2O_2 (concentration $= 8.00 \times 10^{-3}$ mol dm^{-3}). Since the formation of each mole of I_2 requires the reaction of one mole of H_2O_2, it follows that 1.25×10^{-5} moles of the peroxide are used in the formation of 1.25×10^{-5} moles of I_2, which leaves 7.88×10^{-4} moles of unreacted H_2O_2 in the reaction mixture, corresponding to a final concentration of 7.88×10^{-3} mol dm^{-3}. As the concentration of H_2O_2 has fallen by only 1.5 %, it has – to all intents and purposes (and within experimental error) – remained constant during the time taken to generate the blue-black complex:

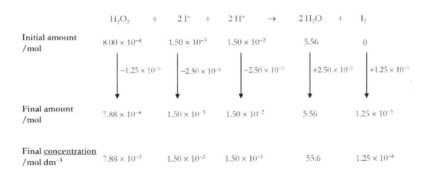

Similarly, the concentrations of the iodide and hydrogen ions – which are higher than that of H_2O_2 – have varied by such a small amount, their

xxx 100 cm^3 of a solution containing $Na_2S_2O_3$ at a concentration of 2.50×10^{-4} mol dm^{-3} contains 2.50×10^{-5} moles of $Na_2S_2O_3$, which in aqueous solution dissociates giving 2.50×10^{-5} moles of $S_2O_3{}^{2-}$ ($Na_2S_2O_3 \rightarrow 2\,Na^+ + S_2O_3{}^{2-}$). If you are not 100 % confident of your grasp of calculations involving concentrations in solution, it is suggested you read *Understanding Calculations in AS and first year A-Level Chemistry: A guide for the utterly confused.*

changes cannot be discerned (when rounded to 3 significant figures). Since the reaction is carried out in aqueous solution, the amount of water generated in the reaction is insignificant compared with that in the bulk solution, which we saw earlier to be present at a concentration of 55.6 mol dm^{-3} (see Footnote[ttt] on page 150). **Table 4.3** shows the results of this investigation.[yyy]

$[I^-] \times 10^3$ /mol dm^{-3}	time taken for solution to turn blue-black (t)/s	$\dfrac{1}{t}$ /s^{-1}
5.00	83	0.012
10.0	41	0.024
15.0	28	0.036
20.0	21	0.048
25.0	17	0.059

Table 4.3   Effect of iodide ion concentration on the rate of the Harcourt-Essen reaction at 25 °C, determined using the iodine-clock method (rate $\propto 1/t$). The initial concentrations of H_2O_2, HNO_3 and $Na_2S_2O_3$ were 8.00×10^{-3}, 0.150 and 2.50×10^{-4} mol dm^{-3}, respectively. See text for further details.

A plot of $1/t$, which is directly proportional to the rate and, therefore, can be used as a proxy ('stand-in') for rate, against $[I^-]$ confirms that the reaction is first-order with respect to the iodide ion (**Figure 4.4**). That is to say, the rate is directly proportional to $[I^-]$ (doubling $[I^-]$, for example, doubles the rate).

[yyy] Notice how the header in the column giving the concentrations of the iodide ion states that the values have been multiplied by 10^3 (a value of 5.00, for example, indicates an iodide concentration of 0.005 mol dm^{-3}).

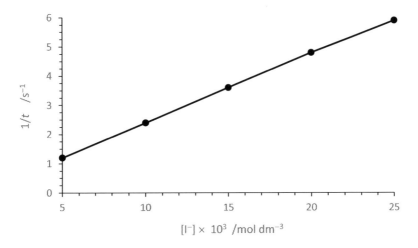

$[I^-] \times 10^3$ /mol dm^{-3}

Figure 4.4   Plot of $1/t$ against $[I^-]$ for the Harcourt-Essen reaction. (Data from **Table 4.3**, where further details are given.)

Although the reaction proceeds by two, independent mechanisms – with rate constants k_2 and k_3 (see above) – the rate equation for each mechanism indicates first-order behaviour in $[I^-]$:

$$\text{rate} \quad = \quad k_2 [H_2O_2][I^-]$$

$$\text{rate} \quad = \quad k_3 [H_2O_2][I^-][H^+]$$

$$\text{overall rate} \quad = \quad k_2 [H_2O_2][I^-] \quad + \quad k_3 [H_2O_2][I^-][H^+]$$

The use of $1/t$ as a 'stand in' for the rate of a reaction provides a simple and convenient means of determining its order with respect to a given reactant. The reaction 'rates' given in **Table 4.3** (as $1/t$ values) are readily converted into conventional reaction rates, with units of mol dm^{-3} s^{-1}, as shown in **Table 4.4**. Take, for example, the values in the first row, where the initial concentration of the iodide ion was 5.00×10^{-3} mol dm^{-3}: it took 83 seconds for 1.25×10^{-5} moles of iodine to be formed (when the blue-black complex appeared, 2nd column). Dividing 1.25×10^{-5} by 83 gives the number of moles of I_2 produced in each second, *i.e.* 1.5×10^{-7} moles (3rd column). The change in the *concentration*

159

of I_2 per second – *i.e.* the rate of the reaction – is obtained by multiplying this figure by 10, giving 1.5×10^{-6} mol dm^{-3} s^{-1}.[zzz] Remember that, because the initial concentrations of the reactants were much higher than that of the thiosulfate ion, we can assume that $[H_2O_2]$, $][I^-]$and $[H^+]$ remained constant over the time during which I_2 was being generated, resulting in a constant rate of reaction throughout the measuring period; the rates given in last column of **Table 4.4** are, effectively, initial rates.

$[I^-] \times 10^3$ /mol dm^{-3}	Time taken to generate 1.25 $\times 10^{-5}$ moles of I_2 /s	Moles of I_2 generated in 1 second /10^{-7} mol	Rate /10^{-6} mol dm^{-3} s^{-1}
5.00	83	1.5	1.5
10.0	41	3.0	3.0
15.0	28	4.5	4.5
20.0	21	6.0	6.0
25.0	17	7.4	7.4

Table 4.4 Calculation of the rate of the Harcourt-Essen reaction, at different concentrations of the iodide ion, from data obtained using the iodine-clock method (taken from **Table 4.3**, where further details are given).

Plotting these initial rates against the (initial) concentration of the iodide ion confirms the reaction is first order in $[I^-]$ (**Figure 4.5**). A similar series of experiments, in which the initial concentration of hydrogen peroxide was varied, showed the reaction to also be first order in H_2O_2: the initial rates shown in **Figure 4.6** were calculated from values of $1/t$ in a manner identical to that described above. As we saw

---

[zzz] The total reaction volume was 100 cm^3, which is one tenth of a cubic decimeter (1 dm^3 = 1000 cm^3). An increase of 1.2×10^{-7} moles in the number of moles of I_2 in a volume 100 cm^3 is the same as an increase of 1.2×10^{-6} moles in 1000 cm^3. (It may help to think of 1 dm^3 as ten, individual cubes of 100 cm^3, each containing 1.2×10^{-7} moles of I_2.)

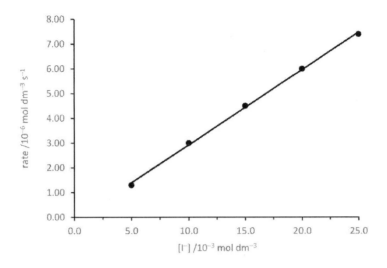

Figure 4.5 Plot of initial rate against [I⁻] for the Harcourt-Essen reaction. (Data from **Table 4.4**, where further details are given.)

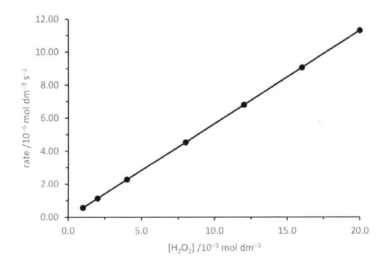

Figure 4.6 Plot of initial rate against [H_2O_2] for the Harcourt-Essen reaction at 25 °C. Rates were determined using the iodine-clock method, as described in the text. The initial concentrations of KI, HNO_3 and $Na_2S_2O_3$ were 5.00×10^{-3}, 0.150 and 2.50×10^{-4} mol dm⁻³, respectively.

above with the iodide ion, this behaviour is reflected in the individual rate equations of the two, separate reaction mechanisms (with rate constants k_2 and k_3).

In contrast to $[H_2O_2]$ and $[I^-]$, experiments in which the initial concentration of the hydrogen ion (from HNO_3) is varied do not reveal such straightforward behaviour. At low concentrations of H^+, the rate is essentially independent of the concentration of the ion (**Figure 4.7**, *upper graph*); in other words, the reaction is zero order in H^+, reflecting the dominance of the mechanism with the second-order rate equation:

$$\text{rate} \quad = \quad k_2[H_2O_2][I^-]$$

It is only at relatively high concentrations of the hydrogen ion that the third-order mechanism begins to make a significant contribution to the overall rate and the reaction 'becomes' first-order in $[H^+]$(**Figure 4.7**, *lower graph*):

$$\text{rate} \quad = \quad k_3[H_2O_2][I^-][H^+]$$

Notice how increasing $[H^+]$ through the lower range of concentrations does, in fact, cause the rate to increase *very* slightly; however, the effect is so small it would not be resolved against the experimental error (**Figure 4.7**, *upper graph*). You may be relieved to know that, in the context of a school laboratory practical or, indeed, an examination question, it is likely that the range of H^+ concentrations will have been chosen to ensure the reaction displays relatively simple kinetics: either zero- or first-order behaviour in $[H^+]$.[aaaa]

As described earlier, where rates were measured by titrating the iodine produced in the reaction (pages 147 – 149), k_2 and k_3 can be obtained by solving simultaneous equations created by substituting the rates measured at any two H^+ concentrations[bbbb] into the rate equation:

$$\text{rate} \quad = \quad k_2[H_2O_2][I^-] \quad + \quad k_3[H_2O_2][I^-][H^+]$$

[aaaa] See, for example, Question 1.1 to 1.7 from the 2018 AQA Paper 3. (Part 1.8 of this question could be answered using a clock method.)

[bbbb] That is to say, any two concentrations of H^+ in the range over which the rate is sensitive to $[H^+]$ (0.100 to 1.00 $mol\ dm^{-3}$ $mol\ dm^{-3}$, in this example).

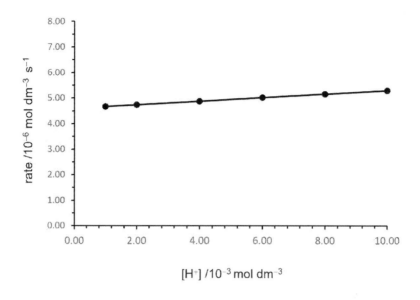

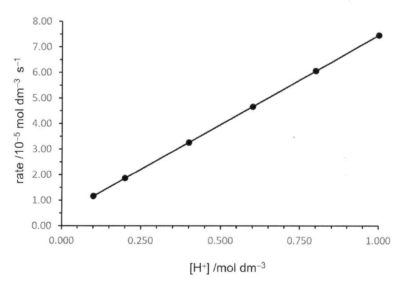

Figure 4.7 Plots of the initial rate of the Harcourt-Essen reaction against [H⁺] over two concentration ranges of the ion. *Upper graph:* 1.00 × 10⁻³ to 10.0 × 10⁻³ mol dm⁻³ mol dm⁻³ H⁺ (as HNO₃); *lower graph:* 0.100 to 1.00 mol dm⁻³ mol dm⁻³ H⁺. Rates were determined using the iodine-clock method, as described in the text. The initial concentrations of H_2O_2, KI and $Na_2S_2O_3$ were 2.00 × 10⁻², 2.00 × 10⁻² and 2.50 × 10⁻⁴ mol dm⁻³, respectively. All reactions were at 25 °C.

At H^+ concentrations of 0.200 and 0.400 mol dm^{-3}, for example, the rates were 1.86×10^{-5} and 3.26×10^{-5} mol dm^{-3} s^{-1} (**Figure 4.7**, *lower graph*), giving:

1.86×10^{-5} mol dm^{-3} s^{-1} =

$\quad$ $k_2 (2.00 \times 10^{-2}$ mol dm$^{-3})(2.00 \times 10^{-2}$ mol dm$^{-3})$ +

$\quad$ $k_3 (2.00 \times 10^{-2}$ mol dm$^{-3})(2.00 \times 10^{-2}$ mol dm$^{-3})$**(0.200 mol dm^{-3})**

and,

3.26×10^{-5} mol dm^{-3} s^{-1} =

$\quad$ $k_2 (2.00 \times 10^{-2}$ mol dm$^{-3})(2.00 \times 10^{-2}$ mol dm$^{-3})$ +

$\quad$ $k_3 (2.00 \times 10^{-2}$ mol dm$^{-3})(2.00 \times 10^{-2}$ mol dm$^{-3})$**(0.400 mol dm^{-3})**

Solving these equations (using the procedure described earlier) gives values for k_2 and k_3 of 0.0115 dm^3 mol^{-1} s^{-1} and 0.175 dm^6 mol^{-2} s^{-1}, respectively. Alternatively, we can obtain k_2 and k_3 using a more accurate (and more elegant) method that is based on the fact that the plot of initial rate against $[H^+]$ gives a straight line (see **Figure 4.7**); in other words, it is of the form $y = mx + c$. To understand why this is the case, it is necessary to remember that, over the period of time leading up to the appearance of the blue-black complex, the concentrations of that H_2O_2, I^- and H^+ remain essentially constant. (As explained above, this is because the reactants are present in large excess over the concentration of the $S_2O_3^{2-}$, which means the generation of enough I_2 to oxidise all of the $S_2O_3^{2-}$ ions results in only a tiny decrease in the concentrations of the reactants.)

Substituting into the rate equation the concentrations of H_2O_2 and I^- employed in the experiment reported in **Figure 4.7** gives:

rate = $\quad$ $k_2 (0.02$ mol dm$^{-3})(0.02$ mol dm$^{-3})$ +

$\quad\quad\quad\quad\quad\quad$ $k_3 (0.02$ mol dm$^{-3})(0.02$ mol dm$^{-3})$ $[H^+]$

As the concentrations of H_2O_2 and I^- were the same in all runs (at all concentrations of H^+), the rate equation simplifies to the following, where $(0.0004 \text{ mol}^2 \text{ dm}^{-6}) \times k_2$ and $(0.0004 \text{ mol}^2 \text{ dm}^{-6}) \times k_3$ are constants and the rate and $[H^+]$ are the only variables:

$$\text{rate} = \underbrace{(0.0004 \text{ mol}^2 \text{ dm}^{-6}) \, k_2}_{a \text{ constant}} + \underbrace{(0.0004 \text{ mol}^2 \text{ dm}^{-6}) \, k_3 \, [H^+]}_{a \text{ constant}}$$

Rearranging the right-hand side puts the equation in the familiar form of that for a straight line:

$$\underset{y}{\text{rate}} = \underbrace{(0.0004 \text{ mol}^2 \text{ dm}^{-6}) \, k_3}_{m} \, \underset{x}{[H^+]} + \underbrace{(0.0004 \text{ mol}^2 \text{ dm}^{-6}) \, k_2}_{c}$$

The gradient (m) of the straight line obtained by plotting rate against $[H^+]$ (**Figure 4.7**, *lower graph*) is $7.00 \times 10^{-5} \text{ s}^{-1}$.[cccc] This constant is equal to $0.0004 \, k_3$:

$$0.0004 \text{ mol}^2 \text{ dm}^{-6} \times k_3 = 7.00 \times 10^{-5} \text{ s}^{-1}$$

giving,

$$k_3 = 0.175 \text{ dm}^6 \text{ mol}^{-2} \text{ s}^{-1}$$

Similarly, extrapolation of the straight line in the lower graph in **Figure 4.7** shows that it intercepts the y-axis at $0.460 \times 10^{-5} \text{ mol dm}^{-3}$; this value corresponds to 'c' in the above expression, therefore:

$$0.0004 \text{ mol}^2 \text{ dm}^{-6} \times k_2 = 0.460 \times 10^{-5} \text{ mol dm}^{-3}$$

giving,

$$k_2 = 0.0115 \text{ dm}^3 \text{ mol}^{-1} \text{ s}^{-1}$$

[cccc] Notice how the units of the gradient have been obtained: $\text{mol dm}^{-3} \text{ s}^{-1}$ (y-axis) divided by mol dm^{-3} (x-axis) gives s^{-1}.

4.3   Measuring an activation energy using the clock method

Clock methods – of which the iodine clock is but one example – then, can serve as relatively simple means of obtaining the kinetic parameters of chemical reactions, even in settings where there is access to only the most basic of laboratory apparatus: most labs have a clock! Clock methods can also be used with the Arrhenius equation to obtain activation energies (and even Arrhenius constants). To illustrate this, we will use the variant of the Harcourt-Essen reaction – described in the previous chapter – in which the peroxodisulfate(VI) ion is used in place of hydrogen peroxide as the oxidising agent:

$$S_2O_8^{2-} \quad + \quad 2\,I^- \quad \rightarrow \quad 2\,SO_4^{2-} \quad + \quad I_2$$

On the basis of the preceding description of the corresponding reaction involving hydrogen peroxide, you might now reasonably expect the iodine produced in this reaction to form the tri-iodide ion:

$$I^- \quad + \quad I_2 \quad \rightleftharpoons \quad I_3^-$$

The I_3^- ion also reacts with peroxodisulfate(VI) in a second-order reaction:

$$S_2O_8^{2-} \quad + \quad 2\,I_3^- \quad \rightarrow \quad 2\,SO_4^{2-} + \quad 3\,I_2$$

The rate constant for the reaction with I_3^-, however, is about half that for the reaction with I^- (the respective values are 1.07×10^{-3} and 2.10×10^{-3} $dm^3\ mol^{-1}\ s^{-1}$). The generation of iodine, and hence the tri-iodide ion, therefore, complicates attempts to determine the rate constant for the oxidation of iodide ions by peroxodisulfate(VI). By using the thiosulfate-clock method, however, this problem is circumvented because the iodine formed in the reaction is rapidly reduced back to iodide ions by the thiosulfate ion, preventing the formation of I_3^- and, thereby, the operation of the competing oxidation.

   In a typical 'run' of the experiment, potassium iodide, sodium thiosulfate and starch are premixed and allowed to settle at the selected reaction temperature in a water bath. The reaction is then initiated by the addition of ammonium peroxodisulfate(VI) solution (at the same temperature) and the timer started.

Table 4.5 presents the results of one such experiment where, for comparison, both $1/t$ and the rate constant are given at each temperature. The reaction times (t) reported in the table were taken from a paper by Moews, Jr and Petrucci,[dddd] who processed these values to obtain the rate constants also given in the table (as described above for the oxidation of iodide by H_2O_2 under acidic conditions).

temperature /K	time taken for solution to turn blue-black (t)/s	$\dfrac{1}{t}$ /s^{-1}	k /dm^3 mol^{-1} s^{-1}
276	189.0	5.3×10^{-3}	1.4×10^{-3}
286	88.0	1.1×10^{-2}	2.9×10^{-3}
297	42.0	2.4×10^{-2}	6.2×10^{-3}
306	21.0	4.8×10^{-2}	1.2×10^{-2}

Table 4.5   Effect of temperature on the rate of oxidation of iodide ions by peroxodisulfate(VI) measured using the iodine-clock method. All reaction mixtures, which were of a final volume of 65 cm^3, contained KI, $(NH_4)_2S_2O_8$ and $Na_2S_2O_3$ at initial concentrations of 0.038, 0.077 and 1.54×10^{-3} mol dm^{-3}, respectively, together with a trace of starch (*ca.* 0.015 %).

The two curves plotted in **Figure 4.8** show how $1/t$ and the rate constant are directly proportional to each other, allowing the use of the former as a proxy for the latter when determining the activation energy from an Arrhenius plot. Moews, Jr and Petrucci plotted ln k against the reciprocal of the temperature ($1/T$), as described here in Chapter 2, to obtain their reported value for the activation energy of 12.4 kcal (51.8 kJ mol^{-1}).

[dddd] P. C. Moews, Jr and R. H. Petrucci (1964) The oxidation of iodide ion by persulfate ion kinetics and mechanism of oxidations by peroxydisulfate. *J. Chem. Ed.* **41**, 549 – 551

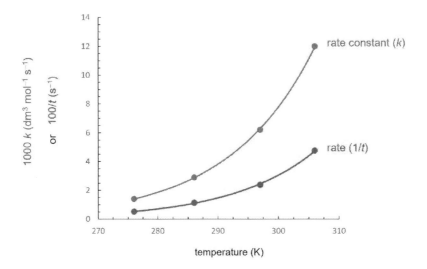

Figure 4.8   Effect of temperature on the rate of oxidation of iodide ions by peroxodisulfate(VI) measured using the iodine-clock method, where the rate is taken as being the reciprocal of the time required for a fixed amount of iodine (5.0×10^{-5} moles[cccc]) to be generated. The corresponding rate constant is also plotted at each temperature. To enable $1/t$ and k to be plotted on the same axes, values have been multiplied by 100 and 1000, respectively. For further details, see **Table 4.5**.

The activation energy can be obtained more simply, however, using $1/t$ as a proxy for the rate constant: although $1/t$ and k are numerically different, they are, as we have seen, directly proportional to each other, so plotting the natural logarithm of either against $1/T$ will give a straight line with the same gradient. This is illustrated in **Figure 4.9**, in which the Arrhenius plots are shown with both $\ln(1/t)$ and $\ln k$ plotted against $1/T$. The gradients of the line-of-best-fit from each plot are very similar and are probably within experimental error: -6120.5 K in the $\ln(1/t)$ plot (*upper graph*) and -6017.5 K in the $\ln k$ plot (*lower graph*).

[cccc] Each reaction contained 1.0×10^{-4} moles of thiosulfate ions (65 cm^3 of 1.54×10^{-3} mol dm^{-3} Na$_2$S$_2$O$_3$). Since each mole of iodine reacts with two moles of thiosulfate, the thiosulfate will be depleted when 5.0×10^{-5} moles of iodine have been formed.

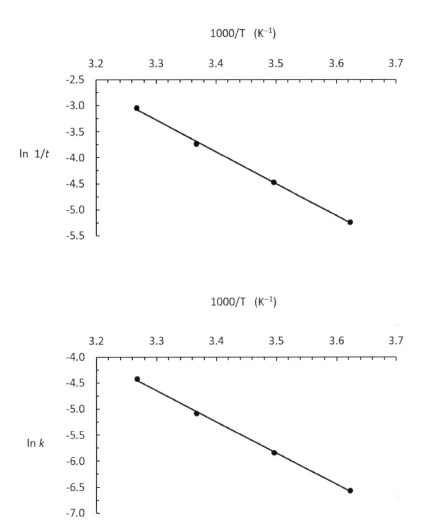

Figure 4.9   Arrhenius plots for the oxidation of iodide ions by peroxodisulfate(VI), measured using the iodine-clock method (see **Table 4.5** and **Figure 4.8** for experimental details). The upper graph was plotted using $1/t$ as a proxy for the rate constant, which was used to obtain the lower graph. Notice how, although the *absolute* values on the y-axes of the two graphs are different, the *change* in the value on the y-axis for a given change in $1/T$ is always the same, hence both plots have the same gradient.

The activation energy calculated using the gradient measured from the plot of ln $(1/t)$ plot is 50.9 kJ mol^{-1}, in close agreement with the value reported by Moews, Jr and Petrucci (51.8 kJ mol^{-1}, page 167):

$$gradient \quad = \quad - \frac{E_a}{R}$$

$$- 6120.5 \text{ K} \quad = \quad - \frac{E_a}{8.314 \text{ J K}^{-1} \text{mol}^{-1}}$$

$$E_a \quad = \quad 6120.5 \text{ K} \times 8.341 \text{ J K}^{-1} \text{mol}^{-1}$$

$$= \quad 50886 \text{ J mol}^{-1}$$

$$= \quad 50.9 \text{ kJ mol}^{-1}$$

Using, instead, the gradient taken from the plot of *ln* k against $1/T$ (*lower graph* in **Figure 4.9**), which has a value of -6017.5, the activation energy is found to be 50029 J mol^{-1} (50.0 kJ mol^{-1}). That this value is slightly lower than the one reported by Moews, Jr and Petrucci probably reflects the more accurate fitting of the line-of-best-fit possible with the aid of modern computer software

Although the two plots shown in **Figure 4.9** have essentially the same gradient (at least within the measurement error achievable in a school laboratory), allowing either to be used to calculate the activation energy, only the data from the lower plot, with ln k on the y-axis, can be used to determine the value of the Arrhenius constant, A. As we saw in Chapter 2 on page 87 (**Figure 2.12**), this entails extrapolating the line-of-best-fit line back to where it intersects the y-axis (where $1/T$ is zero), at which point the value of ln k equals ln A. The line-of-best-fit in the lower graph in **Figure 4.9** intersects the y-axis where ln k is $+15.21$. (Note: it not possible to see this from the figure because the x-axis starts at 3.2×10^{-3} K^{-1}, rather than zero.) Thus,

$$\ln A \quad = \quad 15.21$$

and therefore,

$$A = e^{15.21}$$

$$= 4.03 \times 10^6 \ \text{dm}^3\,\text{mol}^{-1}\,\text{s}^{-1}$$

The value of $\ln(1/t)$ where $1/T$ $(1000/T)$ is zero in the upper graph of **Figure 4.9**, $+16.92$, is not equal to $\ln A$ and therefore cannot be used to obtain A. If $\ln(1/t)$ is being plotted against $1/T$, the logarithmic form of the Arrhenius equation) becomes,

$$\ln\left[\frac{1}{t}\right] = \frac{-E_a}{R} \times \left[\frac{1}{T}\right] + \ln C$$

$$y \quad = \quad m \quad \times \quad x \quad + \quad c$$

where C is just a constant, of value $2.23 \times 10^7 \ \text{dm}^3\,\text{mol}^{-1}\,\text{s}^{-1}$ in this case.

Of the three values for the activation energy stated above – that reported by Moews, Jr and Petrucci; the value of 50886 J mol^{-1} calculated from the plot of $\ln(1/t)$ against $1/T$; and the value of 50029 J mol^{-1} from the corresponding plot of $\ln k$ against $1/T$ – it would appear that the latter is the most accurate. This is because, in combination with the Arrhenius constant of $4.03 \times 10^6 \ \text{dm}^3\,\text{mol}^{-1}\,\text{s}^{-1}$,[ffff] this value gives calculated rate constants (using the Arrhenius equation) for the reaction at the temperatures given in **Table 4.5** closest to the *measured* values reported by Moews, Jr and Petrucci.

4.4   Other 'clocks'

The iodine clock is just one example of a clock method, albeit the version you are most likely to encounter. In Chapter, 1 we explored in some detail the reaction between bromate(V) and bromide ions under acidic conditions:

$$\text{BrO}_3^- + 5\,\text{Br}^- + 6\,\text{H}^+ \rightarrow 3\,\text{Br}_2 + 3\,\text{H}_2\text{O}$$

The rate of this reaction can be measured using the bromine-clock method, in which a fixed amount of phenol and a small amount of the

---

[ffff] This is a relatively small Arrhenius constant, reflecting the fact that the I^- and $\text{S}_2\text{O}_8^{2-}$ ions will react only if they collide in a highly specific orientation.

indicator methyl red (or methyl orange) are included in the reaction mixture. The bromine formed in the reaction reacts rapidly with phenol by electrophilic substitution:

$$C_6H_5OH + Br_2 \rightarrow BrC_6H_4OH + HBr$$

Unlike the corresponding reaction with benzene (C_6H_6), a halogen carrier is not needed as a catalyst when bromine reacts with phenol. This is because a lone pair of electrons from the oxygen atom in phenol delocalises onto the aromatic ring, thereby increasing its electron density and, therefore, reactivity towards electrophiles.

The rapid removal of the Br_2 generated in the reaction between BrO_3^-, Br^- and H^+ ions by its reaction with phenol prevents it from reacting with the indicator. Although the mono-substituted product (BrC_6H_4OH) can undergo further substitution reactions with bromine,[gggg] these are somewhat slower than the initial substitution. Consequently, when all the phenol has undergone mono-substitution, the concentration of bromine rises sharply, reaching a level at which it reacts with the indicator. Upon reaction with bromine, methyl red is bleached. The loss of colour of the indicator occurs quite suddenly, at which point the timer is stopped, giving t, the time taken for a fixed amount of bromine to be formed. As with the iodine-clock method, t is inversely proportional to reaction rate, enabling the determination the order of the reaction with respect to each reactant – and the activation energy – in the usual way (see, for example, Core Practical 14 in the Edexcel course and the article by J. R. Clarke[hhhh]).

4.5   The half-life method

You may have encountered the idea of the half-life in GCSE physics, where it is defined as the time taken for half the radioactive nuclei present in a sample to decay. A similar concept is used in chemical kinetics, where it is particularly useful for identifying first-order behaviour: quite simply, if the half-life of a particular reactant – the time

---

[gggg]Students following the OCR and Edexcel courses will be familiar with the white precipitate of 2,4,6-tribromophenol that is formed when bromine water is added to phenol (along with the observed decolourisation of the bromine water).

[hhhh]J. R. Clarke (1970) The kinetics of the bromate-bromide reaction. *Journal of Chemical Education* **47**, 775 – 778)

taken for its concentration to fall to half its value – is constant, the reaction is first order with respect to the reactant.

In the previous chapter we saw that, although the chemical equation for the decomposition of hydrogen peroxide is usually written to show two molecules of the peroxide reacting,

$$2\,H_2O_2 \quad \rightarrow \quad 2\,H_2O \quad + \quad O_2$$

the reaction is first order with respect to H_2O_2. The time course of hydrogen peroxide consumption in this reaction, shown in **Figure 3.3** in Chapter 3, has been reproduced in **Figure 4.10**, but with the first three half-lives indicated: the time taken for the H_2O_2 concentration to fall from its initial value of 0.050 mol dm^{-3} to 0.025 mol dm^{-3} is 0.592 s; it then takes another 0.592 s for the concentration to halve again to 0.013 mol dm^{-3} and, similarly, a further 0.592 s to halve yet again.

The finding that the half-life of the peroxide is constant confirms that the reaction is first order in H_2O_2 and – as H_2O_2 is the only reactant – the overall reaction is first order. Although the first three, consecutive half-lives were measured in this example, half-lives starting from any concentration of a reactant can be used. We could, for example, have measured the time taken for the concentration of H_2O_2 to fall from 0.040 to 0.020 mol dm^{-3} or 0.030 to 0.015 mol dm^{-3}; it would still have been 0.592 s. When using this method in an examination question to prove that a reaction is first order, you must measure at least three half-lives to confirm the value is constant.

Having measured the half-life, and confirmed that it is constant, it is then a very simple matter to calculate the rate constant using the expression (where $t_{1/2}$ is the half-life):

$$k \quad = \quad \frac{\ln 2}{t_{1/2}}$$

As the natural logarithm (ln) of 2 is 0.693 we could simply write,

$$k \quad = \quad \frac{0.693}{t_{1/2}}$$

but it is perhaps easier to remember the expression containing ln 2. Substituting in the half-live for the decomposition of hydrogen peroxide, gives the first-order rate constant:

173

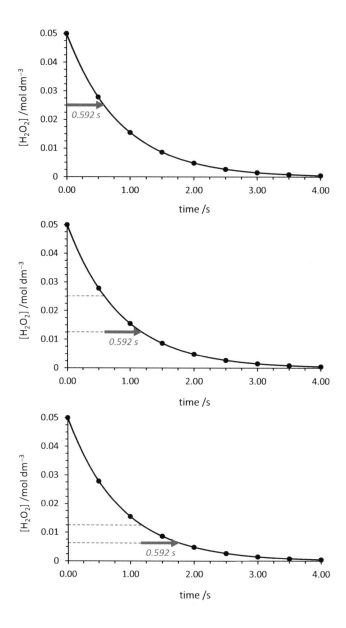

Figure 4.10   Time course of hydrogen peroxide decomposition at 400 °C
and 2.45 × 10⁷ Pa, showing the measurement of the first
three consecutive half-lives (0.592 s): first (*top graph*), second
(*middle graph*) and third (*bottom graph*). The data points were
calculated using kinetic parameters for the reaction taken
from a study by Eric Croiset, Steven F. Rice and Russell G.
Hanush (*AIChE Journal*, 1997, **43**, 2343 – 2352).

174

$$k \quad = \quad \frac{0.693}{0.592 \text{ s}}$$

$$= \quad 1.17 \text{ s}^{-1}$$

Notice how 'processing' the units of the half-life (seconds) through this calculation gives rise to the correct units of the first-order rate constant. Also, bear in mind that this relationship between the half-life and rate constant applies only for first-order reactions, therefore you must first confirm that the reaction is, indeed, first-order by measuring at least three half-lives.

Using the half-life to find the rate constant of a first-order reaction is far more reliable than trying to draw an accurate tangent, especially if the tangent is drawn at the zero time-point (used to measure the initial rate). If an examination question asks you to measure the initial rate from a plot of the type shown in **Figure 4.10**, then of course you must draw a tangent using your best judgement. If, however, you are simply asked to find the order of the reaction, its rate constant and the rate equation, then measuring a few half-lives may be the best place to start. Do study the mark schemes from past examinations papers: increasingly, you will see recognition and acceptance of the half-life method (often as an 'alternative method') by the main examination boards. The method is referred to in the OCR A Specification, yet in my experience surprisingly few students are aware of it.

4.6   Imposing first-order conditions on a reaction: the alkaline hydrolysis of an ester

Such is the simplicity of the half-life method in obtaining a first-order rate constant, physical chemists often employ reaction conditions that cause reactions that are not first order to behave as such. We have encountered already examples of so-called pseudo first-order conditions, in which the initial concentration of just one of the reactants is very small compared with those of the other reactants. We will now look at how such conditions can be used, in conjunction with the half-life method, to obtain the rate constant of *second*-order reaction by first obtaining it as a pseudo first-order rate constant under conditions that cause the reaction to behave as though it were first order.

175

To illustrate this approach, we will consider the hydrolysis of ethylethanoate using sodium hydroxide, giving sodium ethanoate and ethanol:

$$CH_3COOCH_2CH_3 + NaOH \rightarrow CH_3COONa + CH_3CH_2OH$$

This reaction has been shown to be first order with respect to the concentration of both reactants and is, therefore, second order overall.

The rate equation,

$$\text{rate} = k \, [CH_3COOCH_2CH_3][OH^-]$$

reflects a mechanism in which the nucleophilic attack of the hydroxide ion upon the partially-positive carbon atom in the ester linkage is believed to be the rate-determining step:[iiii]

The tetrahedral intermediate formed in this reaction step then rapidly fragments to the products. Students with the AQA examinations board will recognise this mechanism as nucleophilic addition-elimination, by which vegetable oils and animal fats are hydrolysed under alkaline conditions to give soaps (salts of long-chain carboxylic acids) and glycerol:

[iiii] The rate equation is also consistent with breaking in two of the negatively-charged tetrahedral intermediate, formed by the addition of OH^- to the ester, being the rate-determining step (see the following page). The intermediate is not a reactant, so cannot appear in the rate equation, but its concentration is determined by the concentrations of both OH^- and the ester, thereby accounting for the observed kinetics. The evidence, however, points to the initial reaction between OH^- and the ester as the rate-determining step (see e.g. H. S. Levenson and H. A. Smith, *J. Am. Chem. Soc*, 1940, **62**, 2324 – 2327; and P. Sykes, 1986, *A Guidebook to Mechanism in Organic Chemistry*, Sixth Edition, Prentice Hall).

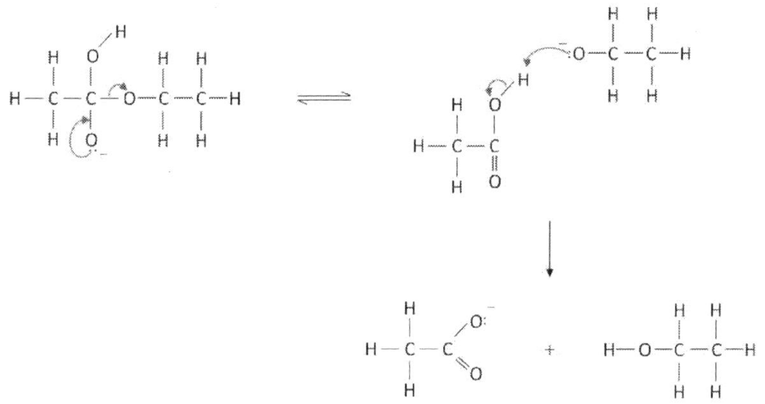

In the graph shown in **Figure 4.11**, the decrease in the concentration of the hydroxide ion has been plotted against time. The three red arrows indicate the time taken for the concentration of the ion to decrease from 0.0500 to 0.0250 mol dm^{-3} (its first half-life), from 0.0250 to 0.0125 mol dm^{-3} (second half-life) and from 0.0125 to 0.00625 mol dm^{-3} (third half-life). Although the reaction is first order with respect to the concentration of the hydroxide ion, the half-life increases over time because the removal of a hydroxide ion requires it to react with the ester, the concentration of which is also decreasing with time. In other words, because the concentrations of both the hydroxide ion and the ester decrease during the course of the reaction, there will be progressively fewer collisions occurring between them with each second that goes by, resulting in an increase in the time taken for the concentration of the reactants to decrease by a given amount. An increase in the half-life of one of the reactants is a characteristic of a reaction that is second order overall.

Now compare **Figure 4.11** with the plot shown in **Figure 4.12**, the difference being that in the latter the initial concentration of the ester has been increased to 0.800 mol dm^{-3} (with that of the hydroxide ion remaining at 0.0500 mol dm^{-3}). Notice that, to capture the higher rate of disappearance of the hydroxide ion, it has been necessary to plot the time in seconds rather than minutes.

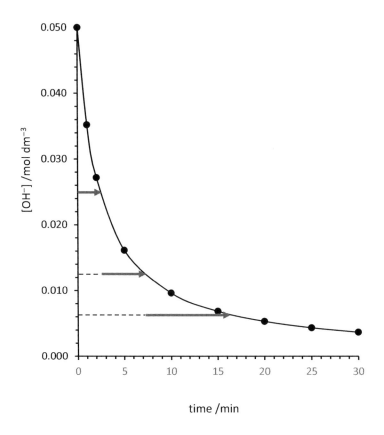

Figure 4.11   Time course of the removal of the hydroxide ion (added as NaOH) in its reaction with ethylethanoate at 25 °C. Both reactants were present at an initial concentration of 0.0500 mol dm^{-3}. The arrows show the first three half-lives of the hydroxide ion (see text for details).

Notice also – and in particular – that the half-life of the OH^{-} ion is now constant (6.2 seconds), indicating that the reaction is behaving as though it were first order overall: first order with respect to the hydroxide ion and *zero order with respect to the ester*. In reality, of course, we know the reaction is first order with respect to the ester. The reason the reaction appears to be first order with respect to the ester is because the change in the concentration of the ester over the course of the reaction is very small – so small, in fact, that it cannot have any effect on the rate.

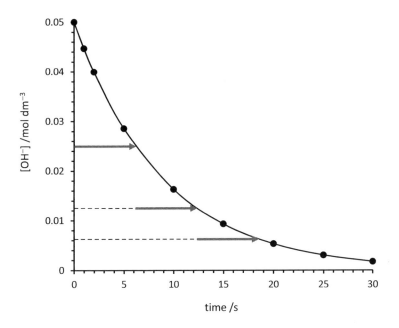

Figure 4.12 Time course of the removal of the hydroxide ion in its reaction with ethylethanoate at 25 °C under pseudo first-order conditions. The initial concentration of the hydroxide ion (as NaOH) was 0.0500 mol dm^{-3}, but that of the ester was 0.800 mol dm^{-3}. The arrows show the first three half-lives of the hydroxide ion are constant (6.2 seconds).

We can understand why there is, effectively, no change in the concentration of the ester by considering the numbers of moles of each reactant at the beginning and end of the reaction. For simplicity, let us assume the total volume of the reaction mixture was 1 dm^3. This means that, at the moment the reaction was initiated, the mixture contained 0.0500 moles of OH$^-$ ions and 0.800 moles of the ester. After 30 seconds, the concentration of OH$^-$ had fallen to 0.00174 mol dm^{-3} (**Figure 4.12**), which means that 0.0483 moles of the ion have reacted (0.0500 minus 0.0483 leaves 0.00174 moles). Since each OH$^-$ ion reacts with one molecule of ethylethanoate, it follows that after 30 seconds 0.0483 moles of the ester have also reacted, leaving 0.752 moles in the reaction mixture (0.800 minus 0.0483). Thus, in the time it has taken for the concentration of OH$^-$ ions to decrease by over 96 %, from 0.0500

179

to 0.00174 mol dm^{-3}, the concentration of the ester has fallen by only 6 % (from 0.800 to 0.752 mol dm^{-3}). Within experimental error, a 6 % decrease in the concentration of the ester is too small to have any impact on the observed reaction rate; therefore, to all intents and purposes, the concentration of ethylethanoate has remained constant. It is as though we had carried out a series of experiments in which the concentration of OH$^-$ had been varied, keeping that of the ester constant, allowing us to observe the effect of the former in isolation.

Look back to Chapter 1, where we considered the second-order nucleophilic reaction of the hydroxide ion and bromoethane: a series of experimental 'runs' was carried out in which the concentration of the OH$^-$ ion was held constant while that of bromoethane was varied (see **Figure 1.8**): by isolating the effects of bromoethane on the rate, we were able to determine that the reaction is first order with respect to this reactant. We have done essentially the same here in the reaction between ethylethanoate and the hydroxide ion, only this time in a single run of the experiment in which the ester was present in excess.

We *could* now draw a series of tangents to the curve in **Figure 4.12**. These would give us the rate at each concentration of the OH$^-$ ion (read-off from the *y*-axis) at an essentially constant concentration of the ester. A plot of rate against OH$^-$ concentration would give a straight line, showing the reaction to be first order with respect to OH$^-$ (see, for example, **Figure 1.9** in Chapter 1 for the corresponding plot for the reaction between bromoethane and OH$^-$). To do so, however, would not only be needlessly time consuming, but would be far less accurate than if we were to analyse the data using the half-life method: the finding that the half-life of the OH$^-$ ion is a constant 6.2 seconds when ethylethanoate is present is in excess (**Figure 4.12**) tells us that the reaction is obeying a first-order rate equation. The *actual* rate equation (confirmed by demonstrating the reaction to also be first order with respect to the ester),

$$\text{rate} \;=\; k \, [CH_3COOCH_2CH_3] \, [OH^-]$$

is of course second order. Since the concentration of the ester is, in effect, fixed at 0.800 mol dm^{-3}, we can write:

$$\text{rate} \;=\; k \times 0.800 \text{ mol dm}^{-3} \times [OH^-]$$

The equation now contains two constants (k, the second-order rate constant and $[CH_3COOCH_2CH_3]$), which we can combine into a single constant, k':

$$k' = k \times 0.800 \text{ mol dm}^{-3}$$

This enables us to write the rate equation with k' in place of $k \times 0.800$ mol dm^{-3}:

$$\text{rate} = k' \, [OH^-]$$

As this is a first-order rate equation, the half-life method may be used to determine the value of k', which is a pseudo first-order rate constant. Using the value of the half-life measured from the plot shown in **Figure 4.12** (6.2 seconds), k' is found to be 0.11 s^{-1}:

$$k' = \frac{\ln 2}{t_{1/2}} = \frac{0.693}{6.2 \text{ s}}$$

$$= 0.11 \text{ s}^{-1}$$

Since,

$$k' = k \times 0.800 \text{ mol dm}^{-3}$$

obtaining k is a simple matter:

$$k = \frac{k'}{0.800 \text{ mol dm}^{-3}}$$

$$= \frac{0.11 \text{ s}^{-1}}{0.800 \text{ mol dm}^{-3}}$$

$$= 0.14 \text{ dm}^3 \text{ mol}^{-1} \text{ s}^{-1}$$

Notice how, by including the appropriate units during the calculation, we arrive at the correct units for the second-order rate constant.

It is because of the relative simplicity and accuracy of the half-life method that physical chemists will often 'impose' first-order conditions on a reaction that may not be first order. Using this approach, we have determined the second-order rate constant for the alkaline hydrolysis of ethylethanoate to be 0.14 dm^3 mol^{-1} s^{-1} at 25 °C.

CHAPTER 4 SUMMARY – the key points

- The rate of a chemical reaction is defined as the change in concentration of a stated reactant or product per unit time. In setting out to investigate the rate of a reaction, the chemist must, therefore, select an observable change – a change which corresponds to the generation of a product or consumption of a reactant – that can readily be measured against time.

- Some of the changes used to determine the rates of chemical reactions can be measured directly, using an appropriate piece of apparatus; these include a change in colour, pH or the evolution of a gas, which can be measured using a colorimeter, a pH meter and a gas syringe, respectively.

- In other situations, the concentration of a reactant or product is measured by carrying out a further chemical reaction – typically, some form of titration. We saw in Chapter 1, for example, how the removal of hydroxide ions during their reactions with halogenoalkanes can be measured by a simple acid-base titration. Reactants or products that are good oxidising agents or reducing agents can be measured by carrying out a redox titration. For example, the iodine generated in the Harcourt-Essen reaction,

$$H_2O_2 \ + \ 2\,I^- \ + \ 2\,H^+ \ \rightarrow \ 2\,H_2O \ + \ I_2$$

can be measured by titration with sodium thiosulfate, using starch as an indicator to show when the last traces of the halogen have been removed in the reaction:

$$I_2 \ + \ 2\,S_2O_3^{2-} \ \rightarrow \ 2\,I^- \ + \ S_4O_6^{2-}$$

- When using a **clock reaction** to measure the rate of a reaction, the reciprocal of the time taken for a fixed amount of product

182

to be generated $(1/t)$ is used as a proxy for the rate. When applied to the Harcourt-Essen reaction, for example, a fixed amount of thiosulfate ions is included in the reaction mixture and the time taken for enough iodine to be generated to oxidise this to tetrathionate ions $(S_4O_6^{2-})$ in the above reaction is measured.

- Not only can the clock method be used to determine the order of a reaction with respect to each reactant (and thereby derive the rate equation), it can be used determine the activation energy, using a 'modified' version of the Arrhenius equation,

$$\ln \left(\frac{1}{t} \right) = \frac{-E_a}{R} \times \left(\frac{1}{T} \right) + \ln C$$

where C is a constant (replacing A, the Arrhenius constant). To obtain the value of A (and, indeed, the rate constant), it is necessary to process the 'raw' $1/t$ values into changes in product concentration per unit time.

- The **half-life** $(t_{1/2})$ of a reaction is the time taken for the concentration of a reactant to fall to half its initial value. The hallmark of a first-order reaction is a constant half-life, therefore the first thing you should ask yourself when presented with a concentration-time plot is whether or not the half-life is constant. If the reaction involves a single reactant, such as the decomposition of hydrogen peroxide,

$$2\,H_2O_2 \;\rightarrow\; 2\,H_2O \;+\; O_2$$

the constant tells you the reaction is first order. The rate constant is then simply calculated using the expression:

$$k = \frac{\ln 2}{t_{1/2}}$$

- The half-life of one of the reactants in a reaction whose order is second, third or greater overall, however, will be constant *if* its concentration is very low compared with the concentrations of the other reactants. Under these so-called **pseudo first-order conditions**, the concentrations of the reactants that

183

are present at relatively high concentration barely change and, therefore, cannot affect the observed rate.

- Such is the simplicity of the half-life method, chemists often carry out reactions that are not first order, under pseudo first-order conditions. From the observed, pseudo first-order rate constant (k'), it is then a simple matter to obtain k – the true, underlying rate constant.

FURTHER READING

Readers taking their studies of reaction kinetics beyond A-level will find more detailed descriptions of the material covered in any of the university-level textbooks on physical chemistry, including *Atkins' Physical Chemistry* (Peter Atkins, Julio de Paula and James Keeler) and *Elements of Physical Chemistry* (Peter Atkins and Julio de Paulo), both of which are excellent and are likely to be found on undergraduate reading lists. Of the textbooks dedicated to the field, Reaction Kinetics (Michael J. Pilling and Paul W. Seakins) is particularly recommended.

For those wishing to expand their knowledge of the reactions described in the text, listed below are some of the scientific papers that were consulted in compiling the text (some of which are also given in footnotes). One or two of these papers are of significant historical interest.

Oxidation of bromide ions by bromate(V) in acidic solution

J. R. Clarke (1970) The kinetics of the bromate-bromide reaction. *J. Chem. Educ.* **47**, 775–778.

C. E. S. Côrtes and R. B. Faria (2001) Revisiting the kinetics and mechanism of the bromate-bromide reaction. *J. Braz. Chem. Soc.* **12**, 775–779.

G. Schmitz (2007) Kinetics of the bromate–bromide reaction at high bromide concentrations. *Int. J. Chem. Kinet.* **39**, 17–21.

Reactions of halogenoalkanes

E. D. Hughes, C. K. Ingold and U. G. Shapiro (1936) Mechanism of substitution at a saturated carbon atom. Part VI. Hydrolysis of *iso*propyl bromide. *J. Chem. Soc.* 225–236.

E. D. Hughes and U. G. Shapiro (1937) Mechanism of substitution at a saturated carbon atom. Part VII. Hydrolysis of *iso*propyl halides. *J. Chem. Soc.* 1177–1183.

E. D. Hughes (1941) Mechanism and kinetics of substitution at a saturated carbon atom. *Trans. Faraday Soc.* **37**, 603–63.

K. A. Cooper, M. L. Dhar, E. D. Hughes, C.K.Ingold, B. J. MacNulty and L. I. Woolf (1948) Mechanism of elimination reactions. Part VII. Solvent effects on rates and product-proportions in uni- and bi-molecular substitution and elimination reactions of alkyl halide and sulphononium slats in hydroxylic solvents. *J. Chem. Soc.* 2043–2049.

V. J. Shiner, Jr, M. J. Boskin and M.L. Smith (1955) The effects of γ-methyl substitution on the rates of the bimolecular displacement and elimination reactions of the alkyl halides. *J. Am. Chem. Soc.* **77**, 5525–5528.

Acid-catalysed halogenation of propanone

G. Archer and R. P. Bell. (1959) Kinetics of the bromination of acetone in concentrated aqueous hydrochloric acid. *J. Chem. Soc.* 3228–3230.

J. E. Dubois and J. Toullec (1973) Cinetique complexe de l'halogenation de cetones en milieu acide – II. Controle par la diffusion des vitesses d'halogenation des enols – consequences. *Tetrahedron* 29, 2859–2866.

J.-E. Dubois, M. El-Alaoul and J. Toullec (1981) Kinetics and thermodynamics of Keto–Enol tautomerism of simple carbonyl compounds: an approach based on a kinetic study of halogenation at low halogen concentrations. *J. Am. Chem. Soc.* **103**, 5393–5401.

Y. Chiang, M. Hojatti, J. R. Keeffe, N. P. Schepp and J. Wirz (1987) Vinyl alcohol: generation and decay kinetics in aqueous solution and determination of the tautomerization equilibrium constant and acid dissociation constants of the aldehyde and enol forms. *J. Am. Chem. Soc.* **109**, 4000–4009.

R. Hochstrasser, A. J. Kresge, N. P. Schepp and J. Wirz (1988) Are the halogenations of simple enols in aqueous solution really diffusion-controlled reactions? *J. Am. Chem. Soc.* **110**, 7875–7876.

Y. Chiang, A. J. Kresge and N. P. Schepp (1989) Temperature coefficients of the rates of acid-catalyzed enolization of acetone and its enol in aqueous and acetonitrile solutions. Comparison of thermodynamic parameters for the Keto–Enol equilibrium in solution with those in the gas phase. *J. Am. Chem. Soc.* **111**, 3977–3980.

J. R. Keeffe, A. J. Kresge and N. P. Schepp (1990) Keto–Enol equilibrium constants of simple monofunctional aldehydes and ketones in aqueous solution. *J. Am. Chem. Soc.* **112**, 4862–4868.

J. P. Birk and D. L. Walters (1992) Three methods for studying the kinetics of the halogenation of acetone. *J. Chem. Educ.* **69**, 585–587.

The Harcourt-Essen reaction

A. V. Harcourt (1867) XLIV.– On the observation of the course of chemical change. *J. Chem. Soc.* **20**, 460–492.

A. V. Harcourt and W. Essen (1867) VII. On the laws of connection between the conditions of a chemical change and its amount. *Phil. Trans. R. Soc.* **157**, 117–137.

W. C. Bray (1932) The mechanism of reactions in aqueous solution. Examples involving equilibria and steady states. *Chem. Rev.* **10**, 161–177.

H. A. Liebhafsky and A. Mohammad (1933) The kinetics of the reduction, in acid solution, of hydrogen peroxide by the iodide ion. *J. Am. Chem. Soc.* **55**, 3977–3986.

H. A. Liebhafsky and A. Mohammad (1934) A third order ionic reaction without appreciable salt effect. *J. Phys. Chem.* **38**, 857–866.

H. F. Shurvell (1967) The kinetics of an ionic reaction. A physical chemistry experiment. *J. Chem. Educ.* **44**, 577–579.

P. D. Sattsangi (2011) A microscale approach to chemical kinetics in the general chemistry laboratory: The potassium iodide hydrogen peroxide iodide-clock reaction. *J. Chem. Educ.* **88**, 184–188.

C. L. Cooper and E. Koubek (1998) A kinetics experiment to demonstrate the role of a catalyst in a chemical reaction. A versatile exercise for general or physical chemistry students. *J. Chem. Educ.* **75**, 87–89.

M. C. Milenković and D. R. Stanisavljev (2012) Role of free radicals in modelling the iodide-peroxide reaction mechanism. *J. Phys. Chem. A* **116**, 5541–5548.

Oxidation of iodide ions by the peroxodisulfate(VI) ion

D. A. House (1962) Kinetics and mechanism of oxidations by peroxydisulfate. *Chem. Rev.* **62**, 185–203.

P. C. Moews, Jr, and R. H. Petrucci (1964) The oxidation of iodide ion by persulfate ion. *J. Chem. Educ.* **41**, 549–551.

A. Indelli and P. L. Bonora (1966) Kinetic study of the reaction of peroxydiphosphate with iodide. *J. Am. Chem. Soc.* **88**, 924–929.

G. S. Laurence and K. J. Ellis (1972) Oxidation of iodide ion by Fe(III) in aqueous solution. *J. Chem. Soc. Dalton Trans.* 2229–2233.

H. F. Shurvell (1966) The activation energy of anionic reaction. A physical chemistry experiment. *J. Chem. Educ.* **43**, 555–556.

Y.-y. Carpenter and H. A. Phillips (2010) Clock reaction: Outreach attraction. *J. Chem. Educ.* **87**, 945–947.

A. E. Burgess and J. C. Davidson (2012) A kinetic–equilibrium study of a triiodide concentration maximum formed by the persulfate–iodide reaction. *J. Chem. Educ.* **89**, 814–816.

Decomposition of hydrogen peroxide

P. A. Giguère and I. D. Liu (1957) Kinetics of the thermal decomposition of hydrogen peroxide vapor. *Can. J. Chem.* **35**, 283–293.

J. Takagi and K. Ishigure (1985) Thermal decomposition of hydrogen peroxide and its effect on reactor water monitoring of boiling water reactors. *Nucl. Sci. Eng.* **89**, 177–186.

C. C. Lin, F. R. Smith, N. Ichikawa, T. Baba and M. Itow (1991) Decomposition of hydrogen peroxide in aqueous solutions at elevated temperatures. *Int. J. Chem. Kinet.* **23**, 971–987.

E. Croiset, S. F. Rice and R. G. Hanush (1997) Hydrogen peroxide decomposition in supercritical water. *AIChE Journal* **43**, 2343–2352.

Z. Hong, A. Farooq, E. A. Barbour, D. F. Davidson and R. K. Hanson (2009) Hydrogen peroxide decomposition rate: a shock tube study using tunable laser absorption of H_2O near 2.5 μm. *J. Phys. Chem. A* **113**, 12929–12925.

D. V. Ilyin, W. A. Goddard III, J. J. Oppenheim and T. Cheng (2019) First-principles-based reaction kinetics from reactive molecular dynamics simulations: application to hydrogen peroxide decomposition. *PNAS* **116**, 18202–18208.

Alkaline hydrolysis of esters

H. Tsujikawa and H. Inoue (1966) The reaction rate of the alkaline hydrolysis of ethyl acetate. *Bull Chem. Soc. Japan* **39**, 1837–1842.

H. S. Levenson and H. A. Smith (1939) Kinetics of the saponification of ethyl esters of normal aliphatic acids. *J. Am. Chem. Soc.* **61**, 1172–1175.

H. S. Levenson and H. A. Smith (1940) Kinetics of the saponification of ethyl esters of several phenyl substituted aliphatic acids. *J. Am. Chem. Soc.* **62**, 2324–2327.

M.-A. Schneider and F. Stoessel (2005) Determination of the kinetic parameters of fast exothermal reactions using a novel microreactor-based calorimeter. *Chem Eng. J.* **115**, 73–78.

K. Das, P. Sahoo, M. Sai Baba, N. Murali and P. Swaminathan (2011) Kinetic studies on saponification of ethyl acetate using an innovative conductivity-monitoring instrument with a pulsating sensor. *Int. J. Chem. Kinet.* **43**, 648–656.

V. C.Eze, A. P. Harvey and A. N. Phan (2015) Determination of the kinetics of biodiesel saponification in alcoholic hydroxide solutions. *Fuel* **140**, 724–730.

Appendix I

HOW TO CONSTRUCT AND COMBINE HALF-EQUATIONS

You can always recognise a half-equation because it includes one or more electrons: if they are on the left-hand side, the equation is for a reduction process (electrons are being added to a species); if they are on the right-hand side, electrons are being removed from a species, so the process is one of oxidation. To decide which species is being reduced or oxidised, you must apply oxidation numbers (also called oxidation states).

Consider the conversion of the dichromate(VI) ion $(Cr_2O_7^{2-})$ to the chromium(III) ion (Cr^{3+}) ion, which takes place when acidified sodium (or potassium) dichromate(VI) is used to oxidise an alcohol to the corresponding carbonyl compound (and, if it is a primary alcohol, through to a carboxylic acid). In this example, the oxidation number of chromium is given within the name of each ion, so we can say that chromium is being reduced from the +6 oxidation state in $Cr_2O_7^{2-}$ to the +3 state in Cr^{3+} (remember, the oxidation number becomes more positive during oxidation and less positive during reduction). We begin constructing the half-equation by balancing the chromium atoms:

$$Cr_2O_7^{2-} \rightarrow 2\,Cr^{3+}$$

To reduce each chromium from +6 to +3 requires the addition of three electrons (changes in oxidation number correspond to numbers of electrons 'gained' or 'lost'). As there are *two* chromium atoms in the dichromate(VI) ion, we must add six electrons to the left-hand side:

$$Cr_2O_7^{2-} + 6\,e^- \rightarrow 2\,Cr^{3+}$$

We now balance the oxygen atoms by adding water molecules. In this case, there are seven oxygen atoms on the left-hand side, so we must add seven H_2O molecules to the right-hand side:

$$Cr_2O_7^{2-} + 6\,e^- \rightarrow 2\,Cr^{3+} + 7\,H_2O$$

Finally, the hydrogen *atoms* are balanced by adding hydrogen *ions*. As there are fourteen hydrogens on the right-hand side, we must add fourteen H^+ ions to the left:

$$Cr_2O_7^{2-} + 6\,e^- + 14\,H^+ \rightarrow 2\,Cr^{3+} + 7\,H_2O$$

You need never get these wrong in the examinations: always check that your final equation is balanced in both atoms *and* charges. (In this example, the total charge on each side is 6+).

From the above example, you can now appreciate why the sodium dichromate(VI) solution is acidified when used to oxidise alcohols. Occasionally, you will be required to construct a half-equation for a reaction taking place under alkaline conditions. In doing so, you first follow the same procedure, exactly as shown above, and then perform a little trick at the end in which you add hydroxide ions. This is best illustrated by taking the reaction of chlorine with cold, dilute sodium hydroxide solution. The reaction is used to produce bleach and is usually covered in the first year at A-level (and is, therefore, covered at AS).

The *overall* reaction (with the omission of state symbols, for clarity) is:

$$Cl_2 + 2\,NaOH \rightarrow NaCl + NaOCl + H_2O$$

It is a disproportionation reaction because the same element (Cl) undergoes both reduction and oxidation (from O in Cl_2 to −1 in NaCl and +1 in NaOCl,[iiii] respectively). Before 'disentangling' the equation into the two, separate half-equations, I will write it as an ionic equation, where I have simply cancelled out the two Na^+ spectator ions on each side:

$$Cl_2 + 2\,OH^- \rightarrow Cl^- + ClO^- + H_2O$$

The half-equation for the *reduction* of chlorine to the chloride ion is simple enough because chlorine is the only element involved:

$$Cl_2 + 2\,e^- \rightarrow 2\,Cl^-$$

Although only *one* of the chlorine atoms – strictly speaking, one *mole* – in the Cl_2 shown in our overall equation is reduced to Cl^-, the half-equation is a stand-alone entity and need only show its reactants

---

[iiii] Sodium chlorate(I) is often written as NaClO but, for the reason given earlier in relation to the corresponding iodine-containing species (see Footnote[ccc], page 117), it is given here as NaOCl. See also the Footnote[m] on page 32 for related bromine-containing species.

and products in the same *ratio* as they are given in the overall equation. It would, however, be equally correct to write:

$$\tfrac{1}{2}\,Cl_2 \;+\; e^- \;\rightarrow\; Cl^-$$

The half-equation for the *oxidation* of chlorine from 0 (in Cl_2) to +1 in the chlorate(I) ion, ClO^-, must however be derived by applying the method shown above. After balancing the chlorine atoms, it is seen to be necessary to remove two electrons (each of the two Cl atoms in Cl_2 is going from 0 to +1, so loses one electron):

$$Cl_2 \;\rightarrow\; 2\,ClO^- \;+\; 2\,e^-$$

We now balance the two oxygens by adding two water molecules:

$$Cl_2 \;+\; 2\,H_2O \;\rightarrow\; 2\,ClO^- \;+\; 2\,e^-$$

Finally, the four hydrogen atoms introduced in the water molecules are balanced by adding four hydrogen ions:

$$Cl_2 \;+\; 2\,H_2O \;\rightarrow\; 2\,ClO^- \;+\; 2\,e^- \;+\; 4\,H^+$$

Now, here comes the little trick. The four hydrogen ions can be made into four water molecules if we give them each a hydroxide ion. We are permitted to add four OH^- ions to the right-hand side, just so long as we do the same to the other side:

$$Cl_2 \;+\; 2\,H_2O \;+\; 4\,OH^- \;\rightarrow\; 2\,ClO^- \;+\; 2\,e^- \;+\; 4\,H^+ \;+\; 4\,OH^-$$

As $H_2O = H^+ | OH^-$, this amounts to:

$$Cl_2 \;+\; 2\,H_2O \;+\; 4\,OH^- \;\rightarrow\; 2\,ClO^- \;+\; 2\,e^- \;+\; 4\,H_2O$$

We now have a half-equation for the oxidation of chlorine to the chlorate(I) ion in alkali. Adding this equation to that for the reduction of chlorine to the chloride ion (underlined and in bold), results in:

$$\underline{\mathbf{Cl_2 \;+\; 2\,e^-}} \;+\; Cl_2 \;+\; 2\,H_2O \;+\; 4\,OH^- \;\rightarrow$$

$$\underline{\mathbf{2\,Cl^-}} \;+\; 2\,ClO^- \;+\; 2\,e^- \;+\; 4\,H_2O$$

Cancelling out the electrons and dividing through by two gives the ionic equation for the overall disproportionation:

$$Cl_2 \;+\; 2\,OH^- \;\rightarrow\; Cl^- \;+\; ClO^- \;+\; H_2O$$

It is then a simple matter to add some spectator ions (in this case, Na^+ because the alkali happens to be sodium hydroxide):

$$Cl_2 + 2\,NaOH \rightarrow NaCl + NaOCl + H_2O$$

Appendix II

HOW CAN A FIRST-ORDER REACTION BE COMPATIBLE WITH COLLISION THEORY?

When considering the isomerisation of cyclopropane to propene in Chapter 3,

we were confronted with a dilemma: how can cyclopropane molecules collide with each other – which is necessary for them to acquire the minimum amount of energy needed for the reaction to take place (the activation energy) – and yet the reaction display first-order kinetics? As the minimum number of molecules involved in a collision is two, we would expect the reaction to be second order with respect to cyclopropane, yet the reaction obeys the first-order rate equation:

$$\text{rate} \;=\; k\,[\textit{cyclo}\text{-}C_3H_6]$$

Not all reactions involving a single reactant, of course, are first order. We saw, for example, that the decomposition of ethanal to methane and carbon monoxide,

$$CH_3CHO \;\rightarrow\; CH_4 \;+\; CO$$

is second order, with the rate equation:

$$\text{rate} \;=\; k\,[CH_3CHO]^2$$

In the 1920s, Frederick Lindemann proposed a model that explains how first-order kinetics is compatible with collision theory. The Lindemann model was subsequently refined by Hinshelwood and has been developed further by several others,[kkkk] but here we will confine

---

[kkkk] For an excellent review, see: F. Di Giacomo (2015) A short account of RRKM theory of unimolecular reactions and of Marcus theory of electron transfer in a historical perspective. *Journal of Chemical Education* **92**, 476 – 481.

our discussion to those elements of the model needed to explain only the observation of first-order behaviour.

Using the isomerisation of cyclopropane to pentene as an example with which to describe the model, we will assume the events leading to the reaction begin with a collision between two molecules. In this collision, one of the cyclopropane molecules becomes energised. In other words, it leaves the collision with more energy than it had when it went into the collision. (Since energy is conserved, it follows that the other molecule loses energy in the collision.)

Representing the energised molecule as $(cyclo\text{-}C_3H_6)^*$, we can show this collision as:

$$cyclo\text{-}C_3H_6 \;+\; cyclo\text{-}C_3H_6 \;\;\rightarrow\;\; (cyclo\text{-}C_3H_6)^* \;+\; cyclo\text{-}C_3H_6$$

The rate of formation of energised cyclopropane molecules is given by,

$$\text{rate of formation of } (cyclo\text{-}C_3H_6)^* \;=\; k_a[cyclo\text{-}C_3H_6][cyclo\text{-}C_3H_6]$$

where k_a is the second-order rate constant for the <u>a</u>ctivation of cyclopropane.

Although the activated species $(cyclo\text{-}C_3H_6)^*$ has enough energy to undergo isomerisation to propene, the model assumes that before this can happen the energy gained in the collision must first be redistributed within the molecule. That is to say, the extra energy possessed by $(cyclo\text{-}C_3H_6)^*$ is initially distributed evenly across the whole molecule; it must then 'migrate' into the bonds that are required to break for it to form propene. In redistributing its energy in this way, the activated species is forming the transition state, $(cyclo\text{-}C_3H_6)^\ddagger$, giving the sequence:

$$(cyclo\text{-}C_3H_6)^* \;\rightarrow\; (cyclo\text{-}C_3H_6)^\ddagger \;\rightarrow\; \text{propene}$$

The activated species, $(cyclo\text{-}C_3H_6)^*$, does not need to gain more energy for this to happen: it possess the same amount of extra energy as the transition state, only this is not localised (or 'concentrated') in the bonds that must break for the isomerisation reaction to take place. Keeping in mind that the activated species must pass through the transition state to form propene, we can write:

$$\text{rate of formation of propene} \;=\; k_b[(cyclo\text{-}C_3H_6)^*]$$

where k_b is the *first*-order rate constant for the conversion of the activated species into propene.

An essential component of the Lindemann model is that a finite amount of time (albeit miniscule) is required for the transition state to be formed from the activated species. Until the activated species has passed through the transition state and formed propene, there is the possibility of it losing energy in a collision with another molecule of cyclopropane:

$$(\textit{cyclo-}C_3H_6)^* + \textit{cyclo-}C_3H_6 \rightarrow \textit{cyclo-}C_3H_6 + \textit{cyclo-}C_3H_6$$

Unless all of the extra energy possessed by $(\textit{cyclo-}C_3H_6)^*$ is transferred to the other molecule, this will result in the formation of two 'slightly energised' molecules of cyclopropane, neither or which possesses enough energy to form the transition state. The rate equation for this second-order reaction is:

$$\text{rate of deactivation of } (\textit{cyclo-}C_3H_6)^* = k_d[(\textit{cyclo-}C_3H_6)^*][\textit{cyclo-}C_3H_6]$$

where k_d is the second-order rate constant for the <u>d</u>eactivation of cyclopropane.

We have seen how the rate of propene formation is governed by the equation,

$$\text{rate of formation of propene} = k_b[(\textit{cyclo-}C_3H_6)^*]$$

but to be of any use to us this rate equation must be written in terms of the concentration of the reactant, cyclopropane. To simplify the mathematics in such situations, physical chemists often apply the so-called **steady-state approximation**. In doing so, it is assumed that during the course of the reaction (*i.e.* except at its very beginning and end), the concentration of the activated species remains constant. For this to be the case, the rate at which $(\textit{cyclo-}C_3H_6)^*$ is formed must equal the rate at which it is being removed. As we have seen, the species is formed in only one reaction, described by the equation,

$$\text{rate of formation of } (\textit{cyclo-}C_3H_6)^* = k_a[\textit{cyclo-}C_3H_6][\textit{cyclo-}C_3H_6]$$

but it removed in *two* reactions: one involves its conversion to propene, the **rate of formation of which equals the rate of removal of** $(\textit{cyclo-}C_3H_6)^*$,

$$\text{rate of formation of propene} \quad = \quad k_b[(cyclo\text{-}C_3H_6)^*]$$

and the other involves its deactivation in a collision with another molecule cyclopropane:

$$\text{rate of deactivation of } (cyclo\text{-}C_3H_6)^* \quad = \quad k_d[(cyclo\text{-}C_3H_6)^*][cyclo\text{-}C_3H_6]$$

For the concentration of $(cyclo\text{-}C_3H_6)^*$ to remain constant, the following must apply:

$$\text{rate of formation of } (cyclo\text{-}C_3H_6)^* \quad =$$

$$\text{rate of deactivation of } (cyclo\text{-}C_3H_6)^* \quad + \quad \text{rate of formation of propene}$$

Therefore,

$$k_a[cyclo\text{-}C_3H_6][cyclo\text{-}C_3H_6] \quad =$$

$$k_d[(cyclo\text{-}C_3H_6)^*][cyclo\text{-}C_3H_6] \quad + \quad k_b[(cyclo\text{-}C_3H_6)^*]$$

Solving this expression for $[(cyclo\text{-}C_3H_6)^*]$ gives:

$$[(cyclo\text{-}C_3H_6)^*] \quad = \quad \frac{k_a[cyclo\text{-}C_3H_6][cyclo\text{-}C_3H_6]}{k_b + k_d[cyclo\text{-}C_3H_6]}$$

The rate equation for the conversion of the activated species into propene can now be written with this solution substituting for $[(cyclo\text{-}C_3H_6)^*]$, thus,

$$\text{rate of formation of propene} \quad = \quad k_b \times [(cyclo\text{-}C_3H_6)^*]$$

becomes,

$$\text{rate of formation of propene} \quad = \quad k_b \times \frac{k_a[cyclo\text{-}C_3H_6][cyclo\text{-}C_3H_6]}{k_b + k_d[cyclo\text{-}C_3H_6]}$$

Now, according to the Lindemann model, reactions that display first-order behaviour do so because the rate at which the activated species passes through the transition state, forming the product, is much slower than the rate at which it undergoes deactivation by colliding with another reactant molecule. In the case of cyclopropane isomerisation to propene, this may be written as,

$$k_b[(\textit{cyclo-}C_3H_6)^*] \quad << \quad k_d[(\textit{cyclo-}C_3H_6)^*][\textit{cyclo-}C_3H_6]$$

which, after cancelling out $[(\textit{cyclo-}C_3H_6)^*]$ from both sides, gives:

$$k_b \quad << \quad k_d[\textit{cyclo-}C_3H_6]$$

As k_b is very, very small compared with $k_d[\textit{cyclo-}C_3H_6]$, we can ignore it in the denominator in the expression derived above for the rate of formation of propene (adding k_b to $k_d[\textit{cyclo-}C_3H_6]$ has no effect). This allows us to then cancel down $[\textit{cyclo-}C_3H_6]$:

$$\text{rate of formation of propene} \quad = \quad k_b \times \frac{k_a[\textit{cyclo-}C_3H_6][\cancel{\textit{cyclo-}C_3H_6}]}{\cancel{k_b} + k_d[\textit{cyclo-}\cancel{C_3H_6}]}$$

If we now group together the three rate constants, we have:

$$\text{rate of formation of propene} \quad = \quad \frac{k_b \times k_a}{k_d} \quad \times \quad [\textit{cyclo-}C_3H_6]$$

As k_b, k_a and k_d are constants, the term on the left is simply another constant, k, thereby reducing the expression to a simple first-order rate equation and, therefore, accounting for the observation of first-order kinetics:

$$\text{rate of formation of propene} \quad = \quad k \times [(\textit{cyclo-}C_3H_6)]$$

In reactions where the rate at which the activated species forms the product *exceeds* the rate of its deactivation through a collision with another reactant molecule, second-order behaviour is observed. Thus, in the decomposition of ethanal,

$$CH_3CHO \rightarrow CH_4 + CO$$

the activated species, $(CH_3CHO)^*$, breaks apart into methane and carbon monoxide before it can be deactivated by collision with another ethanal molecule. The rate of this reaction is limited, then, only by the rate of at which $(CH_3CHO)^*$ can be formed in collisions between *pairs* of CH_3CHO molecules, which is why the reaction is second order.

The Lindemann model was formulated to explain the observation of first-order kinetics in gas-phase reactions. Subsequently, models have been applied to reactions that take place in solution, where the involvement of solvent molecules (such as water) in collision processes must also be considered. We have also seen, for example, in Chapter 3 that nucleophilic substitution in tertiary halogenoalkanes occurs by a first-order process.

Although the account of the Lindemann model given here showed the activated species being formed in a collision between two reactant molecules (cyclopropane), activation can also involve collisions between the reactant and other molecules present, such as an unreactive gas, used to dilute the reactant. In the case of hydrogen peroxide decomposition, which is also a first-order reaction (Chapter 3), activation can involve collisions between the peroxide and water molecules, whether in the liquid phase or under supercritical conditions.

Appendix III

SELECTED EXAMINATION QUESTIONS DISCUSSED

The questions described below, together with their mark schemes and examiners' reports, are openly available on the websites of the respective examination boards. They are all from the summer examination series. The list is by no means exhaustive but gives a good cross-section of the longer questions that draw upon the material covered in the main text. Do ensure you also study the multiple-choice questions; invariably, these contain traps set for the unwary!

By reviewing these questions, you will gain an insight into the extent to which the examiners will call upon you to apply your knowledge of the material in various – perhaps unfamiliar – situations. Some of the questions were selected because they ask you to suggest a mechanism for a reaction that it is unlikely you will have been taught in any detail, such as the oxidation of I^- ions by Fe^{3+} ions or the decomposition of H_2O_2. Although it is always possible for a capable student to arrive at a plausible mechanism (one that is consistent with the rate equation), a student who is already familiar with the mechanism (*e.g.* from reading this book) will be at a distinct advantage.

The questions have been grouped according to examination board for convenience, but do not confine your studies to a single board. A reaction used by one examination board may well appear in a future question set by another. By including the three main boards, I have been able to select questions that I consider to be the most informative – and the most worthy of study – on such important topics as the oxidation of bromide ions by the bromate(V) ion in acid, the Harcourt-Essen reaction, nucleophilic substitution, the iodination of propanone, the decomposition of hydrogen peroxide and the various clock methods. These topics appear time and time again in the examinations.

If nothing else, it is hoped the examination questions below provide justification for the extent and depth to which I have covered the material.

AQA Examinations Board

2019 Paper 2, Question 4

This question concerns a hypothetical reaction, between reactants **P** and **Q**. For reactions with various initial concentrations of the two reactants, the initial concentration of **P** and its concentration after 5 seconds are given. Candidates are asked to calculate the 'initial' rates of the reaction. To do this, of course, you divide the change in the concentration of **P** during the first 5 seconds by 5. In doing so, you are making the assumption that the rate is constant over the time interval: you are, in fact, calculating the *average* rate over 5 seconds (as shown in Chapter 1 for the graphs in **Figure 1.2**, page 6, where progressively smaller time periods were used).

Candidates are then asked to use their initial rates (at different reactant concentrations) to find the order of the reaction with respect to **P** and **Q**, followed by the rate constant.

The final part of this question concerns another hypothetical reaction (between **R** and **S**). Candidates are given the rate at a set of reactant concentrations, told the reaction is second order with respect to both reactants, and asked to calculate the rate constant. This amounts to nothing more than solving the rate equation,

$$\text{rate} \ = \ k \, [R]^2[S]^2$$

for k. This whole question is extraordinary in its simplicity.

2019 Paper 2, Question 5

Another disturbingly simple question. Candidates are given the Arrhenius equation together with the rate constant (at 25 °C) and the activation energy of an unspecified reaction, from which they are required to calculate the Arrhenius constant. Apart from having to convert the temperature into kelvins (and the units of activation energy from $kJ \, mol^{-1}$ to $J \, mol^{-1}$), the question requires absolutely no knowledge of chemistry and could be answered by any half-competent A-level mathematician. According to the mark scheme, the question contained a typographical error that affected some students' ability to answer it, therefore all students were given the full 4 marks.

2018 Paper 2, Question 5

The first part of this question concerns the oxidation of bromide ions by the bromate(V) ion under acidic conditions, covered in Chapter 1:

$$BrO_3^- \ + \ 5\,Br^- \ + \ 6\,H^+ \ \rightarrow \ 3\,Br_2 \ + \ 3\,H_2O$$

Incomplete initial rates data and the rate equation are given, from which candidates are asked to calculate the missing rates and the rate constant.

In the second part of the question, which is perhaps more difficult, candidates are required to obtain the activation energy from the time, t, taken for a fixed amount of bromine to be generated at different temperatures. Although the question is not concerned with methodological details, this is an example of a bromine clock (see other 'clocks', page 171). A logarithmic form of the Arrhenius is given in the question, but, because $1/t$ is being used as a proxy for the rate constant, the Arrhenius constant in the equation has been replaced by another constant (as explained herein on page 171).

2018 Paper 3, Question 1

This question is on the Harcourt-Essen reaction, for which the third-order rate equation is given (see page 117-118):

$$rate \ = \ k\,[H_2O_2][I^-][H^+]$$

It begins by asking candidates why the reaction obeys the first-order rate equation,

$$rate \ = \ k_1\,[H^+]$$

when H_2O_2 and H^+ are present in excess.

This is an example of the 'imposition' of first-order conditions on a reaction, with k_1 being a pseudo first-order rate constant.

According to the current edition of the textbook *AQA Chemistry*, published by Oxford University Press, the iodine clock reaction is an 'Extension feature', meaning it is 'beyond the specification'. Nevertheless, the second part of this question is, essentially, the iodine clock method 'in disguise'. Admittedly, it is possible to answer the question without a knowledge of clock methods, but students 'in the know' will recognise reagent 'X' as playing the same role as the thiosulfate ion and E as iodine (albeit, without the starch).

2017 Paper 2, Question 3

This question begins by asking candidates to calculate a rate constant for a reaction for which the rate equation and some initial rates are given at 25 °C. They are then asked to use the rate constant (and temperature) to calculate the activation energy, using the logarithmic version of the Arrhenius equation, which is provided in the question.

2015, Specimen Paper 2, Question 1

This question begins by asking candidates to deduce a rate equation from initial-rates data. They are also required to calculate an initial rate and the rate constant for a different reaction. Candidates are then asked to explain why doubling the temperature has a far greater effect on the rate of a reaction (for which a first-order rate equation is given) than doubling the concentration of the reactant. The answer to this is embodied within the Arrhenius equation: doubling the concentration causes the collision frequency to double, but doubling the temperature causes a *much greater* increase in the proportion of collisions that occur with an energy of at least the activation energy, as given by the $e^{-(Ea/RT)}$ term. The question does not address the 'dilemma' in accounting for the requirement for collisions in a reaction that displays first-order kinetics (see Appendix II).

The question concludes by asking candidates to calculate the activation energy of a reaction using the logarithmic version of the Arrhenius equation (given in the question) and the necessary data.

EDEXCEL EXAMINATIONS BOARD

2019 Paper 2, Question 7

This question will be looked at here in some detail because it is a relatively simple example of a type of question for which there is an alternative method to find the answer – a method that is often easier than the one the examiners might be expecting you to use.

Candidates are told that the rate of the reaction between nitrogen monoxide and chlorine,[IIII]

$$2\,NO\ +\ Cl_2\ \rightarrow\ 2\,NOCl$$

is 1.09×10^{-2} mol dm^{-3} s^{-1} when the concentration of NO is 0.122 mol dm^{-3} and that of Cl_2 is 0.241 mol dm^{-3}. The rate equation,

$$\text{rate}\ =\ k\,[NO]^2[Cl_2]$$

is given, from which candidates are asked to calculate the concentration of NO that results in a rate of 8.72×10^{-2} mol dm^{-3} s^{-1} if the concentration of Cl_2 is 0.482 mol dm^{-3}.

Increasing $[Cl_2]$ from 0.241 to 0.482 mol dm^{-3} is a 2-fold increase. As the reaction is first order with respect to Cl_2, this *on its own* will cause the rate to increase 2-fold, to 2.18×10^{-2} mol dm^{-3} s^{-1}. The rate has, however, increased to 8.72×10^{-2} mol dm^{-3} s^{-1}, which is four times 2.18×10^{-2} mol dm^{-3} s^{-1}. This means that, whatever the change in [NO] that has been made, it alone is responsible for causing a 4-fold increase in the rate. The reaction, however, is second order in NO, so to bring about a 4-fold increase in the rate, [NO] needs to be increased only 2-fold ($4 = 2^2$). We can conclude, therefore, that the concentration of NO required to result in a rate of 8.72×10^{-2} mol dm^{-3} s^{-1}, when the concentration of Cl_2 is 0.482 mol dm^{-3}, is 0.244 mol dm^{-3}. For further clarification on predicting the effects concentration changes on reaction rates, look back to page 46, where it was shown how it can be helpful to *think of the rate constant as the rate of the reaction when the concentrations of all the reactants equals 1.00 mol dm^{-3}*.

Candidates are then asked to perform a similar calculation to find the concentration of Cl_2 when given the rate at a different concentration of NO. In the next part of the question, candidates are asked to use the data from the first experiment – for which the rate, [NO] and $[Cl_2]$ are given – to find the rate constant. It is, of course, very easy to find k by substituting these values into the rate equation, so why not answer this part of the question first and then use the rate equation, with the rate constant you have calculated, to find the concentrations of NO and Cl_2 required in the first part of the question? This is a much simpler

---

[IIII] This reaction is reversible: see page 76, where we encountered the decomposition of NOCl.

approach, particularly in some of the more difficult questions of this type.

2019 Paper 2, Question 8

Section (b) of this question is a straightforward determination of the activation energy from an Arrhenius plot. A logarithmic form of the Arrhenius equation is provided (already in the form $y = mx + c$), so this is a good question with which to test your skills in this area. Note that, because this is an Edexcel paper, the Arrhenius constant is replaced by 'a constant' (see page 83, above).

2018 Paper 2, Question 9

In this question, the iodine-clock method (page 155) is used initially to measure the rate of the reaction:

$$IO_3^- + 5\,I^- + 6\,H^+ \rightarrow 3\,I_2 + 3\,H_2O$$

Candidates are asked, amongst other questions, to deduce the order of the reaction with respect to the iodate(V) ion. This reaction is very similar to the Harcourt-Essen reaction and its variant using peroxodisulfate(VI) ions as the oxidant and, indeed, in part (c) of the question candidates are asked to deduce the rate equation and rate constant for the Harcourt-Essen reaction from concentration-rate data. Recall that there are two mechanisms for the Harcourt-Essen reaction, operating simultaneously, each with its own rate equation and rate constant (pages 116 to 118). In this question, we are told the reaction is zero order with respect to the H^+ ion. As I commented earlier (page 162), it is very unlikely you will encounter an examination question that requires you to recognise the simultaneous operation of both mechanisms.

This question concludes with the determination of an activation energy from an Arrhenius plot of data from another clock experiment.

2017 Paper 2, Question 9

In first part of this question, candidates are asked to use the half-life of substance **P**, which decomposes in a first-order reaction, to calculate the time taken for its mass to decrease by a stated amount. They are then asked to use some initial-rates data to deduce the rate equation and rate constant of a reaction involving reactants $\mathbf{X}$, $\mathbf{Y}$ and $\mathbf{Z}$.

The question then moves on to some reactions involving actual chemicals: candidates are asked to determine the overall order of the reaction,

$$BrO_3^- + 5\,Br^- + 6\,H^+ \rightarrow 3\,Br_2 + 3\,H_2O$$

which we looked at in detail in Chapter 1 (pages 24–39). The equation for the reaction is not given in the question, only the rate equation,

$$\text{rate} = k[BrO_3^-][Br^-][H^+]^2$$

from which we saw on pages 27–28 gives an overall reaction order of 4 (see also page 40). The question then turns to another reaction looked at in some detail in Chapters 1 and 2, namely the reaction of the hydroxide ion with bromoethane. In this question, candidates are asked to plot some ln k against $1/T$ data and use the Arrhenius equation to find the activation energy (which is given on page 54, above, as 89.5 kJ mol^{-1}).

Sample Assessment Materials, Paper 2, Question 6

Part (c) of this question is on the acid-catalysed reaction of iodine with propanone, which is covered in some detail on pages 106 to 116.

Sample Assessment Materials, Paper 3, Question 3

This question is also on the acid-catalysed reaction of iodine with propanone but, being Paper 3, focuses on the practical aspects of reaction kinetics. The equation for the reaction given in the question shows the hydrogen iodide dissociated into ions:

$$CH_3COCH_{3(aq)} + I_{2(aq)} \rightarrow CH_3COCH_2I_{(aq)} + H^+_{(aq)} + I^-_{(aq)}$$

This is appropriate because the reaction is taking place in aqueous solution and HI is a strong acid (just as you might show HCl dissociated into ions; see also page 109).

The method used to measure the rate of the reaction involves the measurement of the remaining (unreacted) iodine by redox titration against the thiosulfate ion: essentially as described on pages 142–145 for the measurement of iodine formation in the Harcourt-Essen reaction. By adding the samples removed from the reaction mixture at regular time intervals to sodium hydrogen carbonate solution, the sulfuric acid used to acidify the reaction mixture is neutralised, thereby effectively stopping the reaction.

In the final part of the question, candidates are asked to evaluate a proposed mechanism, involving four steps. This mechanism is explained on pages 106 to 109, where the detail provided by the additional elementary reactions should aid understanding.

OCR Examinations Board (Chemistry A)

2019 Paper H432/01, Question 21

In this 'extended response' question on the reaction of bromine with propanone in the presence of hydrochloric acid,

$$CH_3COCH_3 \ + \ Br_2 \ \rightarrow \ CH_3COCH_2Br \ + \ HBr$$

candidates are required to derive the rate equation and rate constant from experimental data. Bromine reacts with propanone by the same mechanism as iodine. See the section in Chapter 3 on the acid-catalysed halogenation of propanone (page 106).

2018 Paper H432/01, Question 17

This question concerns the oxidation of iodide ions by Fe^{3+} ions:

$$2\,Fe^{3+} \ + \ 2\,I^- \ \rightarrow \ I_2 + \ 2\,Fe^{2+}$$

The rate equation, rate constant and mechanism of this reaction are given on page 123, where we looked at how Fe^{3+} ions catalyse the oxidation of iodide ions by the peroxodisulfate(V) ion. Candidates are asked to derive the rate equation and a 'possible' two-step mechanism for the reaction. According to the examiners' report, only the more able candidates were able to suggest a suitable two-step mechanism.

In the second part of this question, candidates are asked to obtain the activation energy and pre-exponential factor (Arrhenius constant) of a reaction for which a plot of ln k against $1/T$ is provided.

2017 Paper H432/01, Question 17

Candidates are presented with a graph of hydrogen peroxide concentration against time, reflecting its decomposition by the reaction:

$$2\,H_2O_2 \quad \rightarrow \quad 2\,H_2O \quad + \quad O_2$$

They are asked to determine the initial rate of the reaction, its order with respect to H_2O_2 and the rate constant. I set this question as homework for all my students, but I also ask them to draw tangents at two, non-zero time points, which give the rate at the corresponding H_2O_2 concentrations. I also ask my students to measure three half-lives, which they find to be the same. Both methods lead to the conclusion that the reaction is first order in H_2O_2 and enable the rate constant to be determined (see Chapter 3, pages 129 to132, and Chapter 4, pages 173 to 175). Bear in mind that the rate constant of 1.17 s^{-1} given in Chapters 3 and 4 is for H_2O_2 decomposition at 400 °C and 2.45×10^7 Pa, so is much higher than the value determined in this question (around 7×10^{-4} s^{-1}).

INDEX

decomposition of H_2O_2,
131-132
elimination from
halogenoalkanes, 105
ethanal decomposition, 81-88
ethylethanoate, alkaline
hydrolysis, 181
Harcourt-Essen reaction,
118, 149, 153, 154
reaction of BrO^- with Br^-
in acid, 29
oxidation of I^- by Fe^{3+}, 123
pseudo-first order,
38, 152, 153, 175, 181, 184
reaction of H_2 with ethane, 80
reaction of I_2 with
propanone, 110
temperature, effect on, 79-82
units of, 30-31, 75-76

Rate-determining step,
16, 40, 95, 137

Rate equation, 23, 40, 41
chain-termination of
methyl radicals, 44
decomposition of H_2O_2,
131, 140
elimination from a
secondary halogeno-
alkane (2nd order), 101
ethylethanoate, alkaline
hydrolysis, 180
Harcourt-Essen reaction,
118, 147, 159, 162
nucleophilic substitution,
(1st order) 39, 93
oxidation of I^- by Fe^{3+}, 123
oxidation of I^- by $S_2O_8^{2-}$, 120
reaction of BrO_3^- with Br^-
in acid, 28, 31, 35, 38
reaction of H_2 with ethane, 71
reaction of I_2 with
CH_3COCH_3, 108
reaction of OH^- with a
secondary halogenoalkane
by substitution, 99, 101
reaction of H_2O with a
secondary halogenoalkane
by substitution, 99, 100
reaction of NO with O_2, 75

Rate of reaction
averaged over a time interval,
1-7
examination question on, 201
definition, 10, 40, 182
measurement using
acid-base titration, 1
clock methods,
155-172, 182-183
examination question on,
202-203, 205
bromination of phenol,
171-172
colour change,
26, 141, 182
disappearing 'X' method, 155
initial-rates method, 17-19
examination question,
201, 203
redox titration, 142-144, 182
examination question on, 207
tangent, from,
8, 11, 12, 17, 18,
129-130, 146
reaction of BrO_3^- with Br^-
in acid, 25-26

examination question on, 202, 206

Rate-limiting step,
see rate-determining step

RDS
see rate-determining step

Reaction-profile diagram,
53, 54, 71, 89, 94
$(CH_3)_3CBr$ reaction
with OH^-, 94
CH_3CH_2Br reaction
with OH^-, 54
H_2 reaction with C_2H_4, 71

Redox reaction
BrO_3^- reaction with Br^-
in acid,
24, 25, 31, 38, 171
Harcourt-Essen reaction,
116-117, 141, 182
H_2O_2 decomposition,
127, 132-135, 173, 183

212

Printed in Great Britain
by Amazon